影印版说明

本书是美国 McGraw-Hill Education 公司 2016 年出版的 *Handbook of Civil Engineering Calculations* Third Edition 的影印版，是目前最新的有关土木工程计算的指导手册，提供了与最新的规范和标准相一致的 3 000 多种计算方法，可以帮助读者准确地进行复杂的设计与施工计算。

考虑到使用方便，影印版分为 7 册：
1 结构工程
  Structural Engineering
2 钢筋混凝土与预应力混凝土工程及设计
  Reinforced and Prestressed Concrete Engineering and Design
3 木结构•土力学
  Timber Engineering•Soil Mechanics
4 测绘、路线设计及公路桥梁
  Surveying, Route Design and Highway Bridges
5 流体力学、水泵、管道及水电
  Fluid Mechanics, Pumps, Piping and Hydro Power
6 供水及雨水系统设计•污水处理及控制
  Water-Supply and Storm-Water System Design•Sanitary Wastewater Treatment and Control
7 工程经济
  Engineering Economics

作者泰勒•G•希克斯是一位经验丰富的顾问工程师，曾负责多个工厂的设计及运行，并在多所工程类院校任教和进行国际讲学，出版了 20 多部相关著作。

材料科学与工程图书工作室
  联系电话 0451-86412421
           0451-86414559
  邮   箱 yh_bj@aliyun.com
           xuyaying81823@gmail.com
           zhxh6414559@aliyun.com

影印版

Third Edition

HANDBOOK OF CIVIL ENGINEERING CALCULATIONS
土木工程计算手册

钢筋混凝土与预应力混凝土工程及设计
Reinforced and Prestressed Concrete Engineering and Design

TYLER G. HICKS, P.E.

哈尔滨工业大学出版社
HARBIN INSTITUTE OF TECHNOLOGY PRESS

黑版贸审字08-2016-106号

Tyler G. Hicks
Handbook of Civil Engineering Calculations, Third Edition
ISBN 978-1-25-958685-9
Copyright © 2016 by McGraw-Hill Education.
All rights reserved. No part of this publication may be reproduced or transmitted in any form or by any means, electronic or mechanical, including without limitation photocopying, recording, taping, or any database, information or retrieval system, without the prior written permission of the publisher.

This authorized English reprint edition is jointly published by McGraw-Hill Education and Harbin Institute of Technology Press Co. Ltd.This edition is authorized for sale in the People's Republic of China only, excluding Hong Kong, Macao SAR and Taiwan.
Copyright © 2017 by McGraw-Hill Education.

版权所有。未经出版人事先书面许可，对本出版物的任何部分不得以任何方式或途径复制或传播，包括但不限于复印、录制、录音，或通过任何数据库、信息或可检索的系统。

本授权英文影印版由麦格劳-希尔教育出版公司和哈尔滨工业大学出版社有限公司合作出版。此版本经授权仅限在中华人民共和国境内（不包括香港特别行政区、澳门特别行政区和台湾地区）销售。

版权©2017由麦格劳-希尔教育出版公司所有。

本书封面贴有McGraw-Hill Education公司防伪标签，无标签者不得销售。

图书在版编目（CIP）数据

土木工程计算手册. 钢筋混凝土与预应力混凝土工程及设计=Handbook of Civil Engineering Calculations. Reinforced and Prestressed Concrete Engineering and Design：英文/（美）泰勒·G·希克斯（Tyler G. Hicks）主编.—影印本.—哈尔滨：哈尔滨工业大学出版社，2017.3
ISBN 978-7-5603-6343-1

Ⅰ.①土… Ⅱ.①泰… Ⅲ.①土木工程-工程计算-技术手册-英文②钢筋混凝土-混凝土工程-工程计算-技术手册-英文③预应力混凝土-混凝土工程-工程计算-技术手册-英文 Ⅳ.①TU-32

中国版本图书馆CIP数据核字（2017）第041758号

材料科学与工程
图书工作室

责任编辑	杨 桦 张秀华 许雅莹
出版发行	哈尔滨工业大学出版社
社　　址	哈尔滨市南岗区复华四道街10号 邮编150006
传　　真	0451-86414749
网　　址	http://hitpress.hit.edu.cn
印　　刷	哈尔滨市石桥印务有限公司
开　　本	660mm×980mm 1/16 印张 8
版　　次	2017年3月第1版 2017年3月第1次印刷
书　　号	ISBN 978-7-5603-6343-1
定　　价	40.00元

（如因印刷质量问题影响阅读，我社负责调换）

HANDBOOK OF CIVIL ENGINEERING CALCULATIONS

Tyler G. Hicks, P.E., Editor
International Engineering Associates
Member: American Society of Mechanical Engineers
United States Naval Institute

S. David Hicks, Coordinating Editor

George K. Korley, P.E.
President/CEO
Korley Engineering Consultants LLC
New York, N.Y.
Contributor, LRFD, Section 1

Third Edition

New York Chicago San Francisco Athens London
Madrid Mexico City Milan New Delhi
Singapore Sydney Toronto

CONTENTS

Preface　vii
How to Use This Handbook　xi

第 1 册
Section 1. Structural Engineering　1.1

第 2 册（本册）
Section 2. Reinforced and Prestressed Concrete Engineering and Design　2.1

第 3 册
Section 3. Timber Engineering　3.1

Section 4. Soil Mechanics　4.1

第 4 册
Section 5. Surveying, Route Design, and Highway Bridges　5.1

第 5 册
Section 6. Fluid Mechanics, Pumps, Piping, and Hydro Power　6.1

第 6 册
Section 7. Water-Supply and Storm-Water System Design　7.1

Section 8. Sanitary Wastewater Treatment and Control　8.1

第 7 册
Section 9. Engineering Economics　9.1

Index　I.1

PREFACE

This third edition of this handbook has been thoroughly updated to reflect the new developments in civil engineering since the publication of the second edition.

Thus, Section 1 contains 35 new LRFD (load resistance factor design) calculation procedures. The new procedures show the civil engineer how to use LRFD in his or her daily design work, and since LRFD is a preferred design method among recent civil engineering graduates, the handbook is right in line with current design methods.

However, since an alternative design method is still used by some engineers around the world, namely ASD—Allowable Stress Design—it, too, is covered in this handbook. Thus the civil engineer will find both methods available to him or her so a suitable design method can be chosen using the third edition of this handbook.

Section 2, on reinforced and prestressed concrete, has been updated with new calculation procedures. The calculation procedures given in this handbook provide a wide-ranging coverage of this important specialty in civil engineering.

Section 3, timber engineering, has new methods and procedures for making calculations in this popular branch of civil engineering.

Section 4, soil mechanics, provides new ways to make soil calculations. Since nearly all structures require soil calculations, the new method given is an important addition to this handbook.

Section 5, surveying, route design, and highway bridges, has important new calculation procedures. Highways and roads, in much of the civilized world, are in poor condition. Potholes, cracks, and edge deterioration are rampant. Drivers often suffer blown tires, broken axles, and damaged wheels on such highways and roads.

When complaints are filed, the most frequent answer heard is "Roadway budgets have been cut; we don't have the money for repairs." In the United States, Congress is considering a major highway repair bill, but it seems to be moving very slowly—much slower than it should—considering the poor conditions of highways and local roads.

If the bill passes and money is appropriated for highway and road repairs, an enormous amount of work will be required. Civil engineers will have at least 10 years of highway and road work ahead of them. And, as part of the bill, new highways will be built. This will open opportunities for creative designs that make for safer driving in all types of weather. Truly, an exciting future is open to civil engineers in highway and road work.

As part of highway and road repairs, bridges will also need lots of work, plus—in some cases—complete replacement. Bridges have suffered numerous failures in recent years, with unfortunate loss of life. Many highway bridges need extensive repairs to steel elements, along with reinforced concrete rehabilitation.

Again, money is needed for this important work. And, like highways, the excuse is lack of funding. Hopefully, money assigned to highway repairs will include funds for bridge repair and replacement.

With bridge design constantly improving and new repair methods being developed, highway spans will be upgraded to reduce the loss of life from structural failures. The third edition of this handbook includes much valuable material on safe and effective bridge design. Civil engineers are urged to become familiar with codes covering safe bridge design. Knowing, and using, such codes will enhance almost every design, while making it safer for drivers and pedestrians.

With the highway, road, and bridge work ahead in next one or two decades, civil engineers have a very bright future. These engineers will seldom want for an important position. The demand for their skills will be ongoing for a very long time.

Section 6, fluid mechanics, pumps, piping, and hydro power, has new calculation procedures in it. A new method of generating hydro power in drinking water, irrigation, and wastewater lines is presented in this section. Also, a proven way for preliminary hydro power generating unit selection is given.

Section 7, water supply and storm-water system design, presents new ideas on water supply and usage. Water shortages are plaguing many parts of the world today. And these water shortages go far beyond the drinking water supply. In some countries the water shortage extends to farms. Owners of farms do not have enough natural water to irrigate their crops. So, some farmers move from nearly barren lands to more copiously supplied areas where they can safely grow their crops.

Drinking water supplies, covered in this section of this handbook, are also in short supply in some countries. Such shortages are critical to people's health. Finding sufficient potable water and delivering it to people's homes and work places is a critical task for civil engineers. The design of safe drinking water systems is also presented in this section. Thus, the civil engineer has the tools to provide safe drinking water, regardless of its source.

And since storm waters may be a source of drinking water, and a scourge to roads and buildings, its control is also provided. Full coverage of storm-water runoff rates, rainfall intensity, and the sizing of sewer pipes is also given in this section. Thus, the civil engineer is fully prepared to both supply and control water in all its uses and sources.

Section 8 covers sanitary wastewater treatment and control—an important topic in today's environmentally conscious world. The newest treatment methods are discussed in a number of important calculation procedures.

Section 9 covers engineering economics. This section has been fully revised so it focuses on the key calculations the civil engineer must make during his or her career. The result is a laser-like focus on the important calculation procedures used in daily civil engineering practice.

Since the second edition of this book was published, there have been major changes in civil engineering. These changes include following:

- **Antiterrorism construction** to protect large and small buildings structurally against terror attacks such as those that have occurred in New York, Paris, Mali, etc.
- **Increased building security features** are now included in almost every major—and many minor—structures to which the public has access. The increased security is to prevent internal and external sabotage and terrorism that could endanger the occupants and the structure.
- **Building Codes have been changed** to provide better protection for occupants and structures. These Code changes affect the daily design procedures of many civil engineers. Such changes are reflected in a number of the calculation procedures in this handbook.
- **"Green" building design is more popular than ever.** Buildings now win awards for their "green" efficiency that reduces energy consumption in new, existing, and rehabilitated buildings. Such "green" awards are important leasing sales features.
- **Major steps to improve indoor air quality (IAQ)** for all buildings have been taken in building design. These steps include much more than prohibiting occupants from smoking in buildings. New rules prohibit smoking within stated distances of the exterior of buildings. IAQ is a major concern in office buildings, schools, hotels, motels, factories, and other buildings throughout the world.
- **Better hurricane, tornado, flood, and wave resistance design of buildings** and other public structures after several disasters including Hurricanes Katrina and Sandy, the

superstorms in the Pacific and Indian Ocean. The loss of more than 250,000 lives in such storms has civil engineers searching for better ways to design, and build, structure—buildings, bridges, dams, etc.—to withstand the enormous forces of nature while protecting occupants. Also under study are (a) early-warning systems to alert people to the onset of dangerous conditions and (b) better escape routes for people fleeing affected areas. New approaches to levee and flood wall design, especially in cities like New Orleans, are being used to prevent recurrence of Hurricane Katrina losses. All these changes will be the work of civil engineers, with the assistance of other specialized professionals.

Taller buildings are being constructed in major cities around the world. Thus, 1,000-foot+ (305 m) mixed-use buildings (residences, stores, offices) are being constructed worldwide. Some even have wind turbines to generate the electric power needed to run the building. Civil engineers will be busy designing the foundations, structural members, and reinforced and prestressed concrete elements of these extra-tall buildings for years to come.

And with the emphasis on "clean energy," hydro power sites are being developed at a faster pace than in many previous decades. New, and updated, generating facilities offer more than just electricity. Some serve as water-supply sources for fresh water used in both agricultural and domestic systems. Sites being developed are in areas throughout the world because electricity is needed almost everywhere on the globe. Civil engineers have, and will continue to have, a major role in the design and construction of these new, and updated, hydro power facilities.

With so many changes "on the drawing board" and computer screen, civil engineers and designers are seeking ways to include the advancements in their current and future design of buildings, bridges, dams, and other structures. This third edition includes many of the proposed changes so designers can include them in their thinking and calculations.

While there are computer programs that help the civil engineer with a variety of engineering calculations, such programs are highly specialized and do not have the breadth of coverage this handbook provides. Further, such computer programs are usually expensive. Because of their high cost, these computer programs can be justified only when a civil engineer makes a number of repetitive calculations on almost a daily basis. In contrast, this handbook can be used in the office, field, drafting room, or laboratory. It provides industry-wide coverage in a convenient and affordable package. As such, this handbook fills a long-existing need felt by civil engineers worldwide.

In contrast, civil engineers using civil-engineering computer programs often find data-entry time requirements excessive for quick one-off-type calculations. When one-off-type calculations are needed, most civil engineers today turn to their electronic calculator, desktop, or laptop computer and perform the necessary steps to obtain the solution desired. But where repetitive calculations are required, a purchased computer program will save time and energy in the usual medium-size or large civil-engineering design office. Small civil-engineering offices generally resort to manual calculation for even repetitive procedures because the investment for one or more major calculation programs is difficult to justify in economic terms.

Even when purchased computer programs are extensively used, careful civil engineers still insist on manually checking results on a random basis to be certain the program is accurate. This checking can be speeded by any of the calculation procedures given in this handbook. Many civil engineers remark to the author that they feel safer, knowing they have manually verified the computer results on a spot-check basis. With liability for civil engineering designs extending beyond the lifetime of the designer, every civil engineer seeks the "security blanket" provided by manual verification of the results furnished by a computer program run on a desktop, laptop, or workstation computer. This handbook gives the tools needed for manual verification of some 2000 civil engineering calculation procedures.

Each section in this handbook is written by one or more experienced professional engineers who is a specialist in the field covered. The contributors draw on their wide experience in their field to give each calculation procedure an in-depth coverage of its topic. So the person using the procedure gets step-by-step instructions for making the calculation plus background information on the subject that is the topic of the procedure.

And because the handbook is designed for worldwide use, both earlier and more modern topics are covered. For example, the handbook includes concise coverage of riveted girders, columns, and connections. While today's civil engineer may say that riveted construction is a method long past its prime, there are millions of existing structures worldwide that were built using rivets. So when a civil engineer is called on to expand, rehabilitate, or tear down such a structure, he or she must be able to analyze the riveted portions of the structure. This handbook provides that capability in a convenient and concise form.

In the realm of modern design techniques, the load and resistance factor method (LRFD) is covered with more than 30 calculation procedures showing its use in various design situations. The LRFD method is ultimately expected to replace the well-known and widely used allowable stress design (ASD) method for structural steel building frameworks. In today's design world, many civil engineers are learning the advantages of the LRFD method and growing to prefer it over the ASD method.

Also included in this handbook is a comprehensive section titled "How to Use This Handbook." It details the variety of ways a civil engineer can use this handbook in his or her daily engineering work. Included as part of this section are steps showing the civil engineer how to construct a private list of SI conversion factors for the specific work the engineer specializes in.

The step-by-step *practical* and *applied* calculation procedures in this handbook are arranged so they can be followed by anyone with an engineering or scientific background. Each worked-out procedure presents fully explained and illustrated steps for solving similar problems in civil engineering design, research, field, academic, or license-examination situations. For any applied problem, all the civil engineer needs to do is place his or her calculation sheets alongside this handbook and follow the step-by-step procedure line for line to obtain the desired solution for the actual real-life problem. By following the calculation procedures in this handbook, the civil engineer, scientist, or technician will obtain accurate results in minimum time with least effort. And the approaches and solutions presented are modern throughout.

The editor hopes this handbook is helpful to civil engineers worldwide. If the handbook user finds procedures that belong in the book but have been left out, the editor urges the engineer to send the title of the procedure to him, in care of the publisher. If the procedure is useful, the editor will ask for the entire text. And if the text is publishable, the editor will include the calculation procedure in the next edition of the handbook. Full credit will be given to the person sending the procedure to the editor. And if users find any errors in the handbook, the editor will be grateful for having these called to his attention. Such errors will be corrected in the next printing of the handbook. In closing, the editor hopes that civil engineers worldwide find this handbook helpful in their daily work.

TYLER G. HICKS, P.E.

HOW TO USE THIS HANDBOOK

There are two ways to enter this handbook to obtain the maximum benefit from the time invested. The first entry is through the index; the second is through the table of contents of the section covering the discipline, or related discipline, concerned. Each method is discussed in detail below.

Index. Great care and considerable time were expended on preparation of the index of this handbook so that it would be of maximum use to every reader. As a general guide, enter the index using the generic term for the type of calculation procedure being considered. Thus, for the design of a beam, enter at *beam(s)*. From here, progress to the specific type of beam being considered—such as *continuous, of steel*. Once the page number or numbers of the appropriate calculation procedure are determined, turn to them to find the step-by-step instructions and worked-out example that can be followed to solve the problem quickly and accurately.

Contents. The contents at the beginning of each section lists the titles of the calculation procedures contained in that section. Where extensive use of any section is contemplated, the editor suggests that the reader might benefit from an occasional glance at the table of contents of that section. Such a glance will give the user of this handbook an understanding of the breadth and coverage of a given section, or a series of sections. Then, when he or she turns to this handbook for assistance, the reader will be able more rapidly to find the calculation procedure he or she seeks.

Calculation Procedures. Each calculation procedure is a unit in itself. However, any given calculation procedure will contain subprocedures that might be useful to the reader. Thus, a calculation procedure on pump selection will contain subprocedures on pipe friction loss, pump static and dynamic heads, etc. Should the reader of this handbook wish to make a computation using any of such subprocedures, he or she will find the worked-out steps that are presented both useful and precise. Hence, the handbook contains numerous valuable procedures that are useful in solving a variety of applied civil engineering problems.

One other important point that should be noted about the calculation procedures presented in this handbook is that many of the calculation procedures are equally applicable in a variety of disciplines. Thus, a beam-selection procedure can be used for civil-, chemical-, mechanical-, electrical-, and nuclear-engineering activities, as well as some others. Hence, the reader might consider a temporary neutrality for his or her particular specialty when using the handbook because the calculation procedures are designed for universal use.

Any of the calculation procedures presented can be programmed on a computer. Such programming permits rapid solution of a variety of design problems. With the growing use of low-cost time sharing, more engineering design problems are being solved using a remote terminal in the engineering office. The editor hopes that engineers throughout the world will make greater use of work stations and portable computers in solving applied engineering problems. This modern equipment promises greater speed and accuracy for nearly all the complex design problems that must be solved in today's world of engineering.

To make the calculation procedures more amenable to computer solution (while maintaining ease of solution with a handheld calculator), a number of the algorithms in the handbook have been revised to permit faster programming in a computer environment. This enhances ease of solution for any method used—work station, portable computer, or calculator.

SI Usage. The technical and scientific community throughout the world accepts the SI (System International) for use in both applied and theoretical calculations. With such widespread acceptance of SI, every engineer must become proficient in the use of this system of units if he or she is to remain up-to-date. For this reason, every calculation procedure in this handbook is given in both the United States Customary System (USCS) and SI. This will help all engineers become proficient in using both systems of units. In this handbook the USCS unit is generally given first, followed by the SI value in parentheses or brackets. Thus, if the USCS unit is 10 ft, it will be expressed as 10 ft (3 m).

Engineers accustomed to working in USCS are often timid about using SI. There really aren't any sound reasons for these fears. SI is a logical, easily understood, and readily manipulated group of units. Most engineers grow to prefer SI, once they become familiar with it and overcome their fears. This handbook should do much to "convert" USCS-user engineers to SI because it presents all calculation procedures in both the known and unknown units.

Overseas engineers who must work in USCS because they have a job requiring its usage will find the dual-unit presentation of calculation procedures most helpful. Knowing SI, they can easily convert to USCS because all procedures, tables, and illustrations are presented in dual units.

Learning SI. An efficient way for the USCS-conversant engineer to learn SI follows these steps:

1. List the units of measurement commonly used in your daily work.
2. Insert, opposite each USCS unit, the usual SI unit used; Table 1 shows a variety of commonly used quantities and the corresponding SI units.
3. Find, from a table of conversion factors, such as Table 2, the value to use to convert the USCS unit to SI, and insert it in your list. (Most engineers prefer a conversion factor that can be used as a multiplier of the USCS unit to give the SI unit.)
4. Apply the conversion factors whenever you have an opportunity. Think in terms of SI when you encounter a USCS unit.
5. Recognize—here and now—that the most difficult aspect of SI is becoming comfortable with the names and magnitude of the units. Numerical conversion is simple, once you've set up *your own* conversion table. So think Pascal whenever you encounter pounds per square inch pressure, Newton whenever you deal with a force in pounds, etc.

SI Table for a Civil Engineer. Let's say you're a civil engineer and you wish to construct a conversion table and SI literacy document for yourself. List the units you commonly meet in your daily work; Table 1 is the list compiled by one civil engineer. Next, list the SI unit equivalent for the USCS unit. Obtain the equivalent from Table 2. Then, using Table 2 again, insert the conversion multiplier in Table 1.

Keep Table 1 handy at your desk and add new units to it as you encounter them in your work. Over a period of time you will build a personal conversion table that will be valuable to you whenever you must use SI units. Further, since *you* compiled the table, it will have a familiar and nonfrightening look, which will give you greater confidence in using SI.

TABLE 1. Commonly Used USCS and SI Units*

USCS unit	SI unit	SI symbol	Conversion factor—multiply USCS unit by this factor to obtain the SI unit
square feet	square meters	m^2	0.0929
cubic feet	cubic meters	m^3	0.2831
pounds per square inch	kilopascal	kPa	6.894
pound force	newton	N	4.448
foot pound torque	newton-meter	Nm	1.356
kip-feet	kilo-newton	kNm	1.355
gallons per minute	liters per second	L/s	0.06309
kips per square inch	megapascal	MPa	6.89
inch	millimeter	mm	25.4
feet	millimeter	mm	304.8
	meter	m	0.3048
square inch	square millimeter	mm^2	0.0006452
cubic inch	cubic millimeter	mm^3	0.00001638
$inch^4$	$millimeter^4$	mm^4	0.000000416
pound per cubic foot	kilogram per cubic meter	kg/m^3	16.0
pound per foot	kilogram per meter	kg/m	1.49
pound per foot force	Newton per meter	N/m	14.59
pound per inch force	Newton per meter	N/m	175.1
pound per foot density	kilogram per meter	kg/m	1.488
pound per inch density	kilogram per meter	kg/m	17.86
pound per square inch load concentration	kilogram per square meter	kg/m^2	703.0
pound per square foot load concentration	kilogram per square meter	kg/m^2	4.88
pound per square foot pressure	Pascal	Pa	47.88
inch-pound torque	Newton-meter	N-m	0.1129
chain	meter	m	20.117
fathom	meter	m	1.8288
cubic foot per second	cubic meter per second	m^3/s	0.02831
$foot^4$ (area moment of inertia)	$meter^4$	m^4	0.0086309
mile	meter	m	0.0000254
square mile	square meter	m^2	2589998.0
pound per gallon (UK liquid)	kilogram per cubic meter	kg/m^3	99.77
pound per gallon (U.S. liquid)	kilogram per cubic meter	kg/m^3	119.83
poundal	Newton	N	0.11382
square (100 square feet)	square meter	m^2	9.29
ton (long 2,240 lb)	kilogram	kg	1016.04
ton (short 2,000 lb)	kilogram	kg	907.18
ton, short, per cubic yard	kilogram per cubic meter	kg/m^3	1186.55
ton, long, per cubic yard	kilogram per cubic meter	kg/m^3	1328.93
ton force (2,000 lbf)	Newton	N	8896.44
yard, length	meter	m	0.0914

(continued)

TABLE 1. Commonly Used USCS and SI Units* (*Continued*)

USCS unit	SI unit	SI symbol	Conversion factor—multiply USCS unit by this factor to obtain the SI unit
square yard	square meter	m^2	0.08361
cubic yard	cubic meter	m^3	0.076455
acre feet	cubic meter	m^3	1233.49
acre	square meter	m^2	4046.87
cubic foot per minute	cubic meter per second	m^3/s	0.0004719

*Because of space limitations this table is abbreviated. For a typical engineering practice an actual table would be many times this length.

TABLE 2. Typical Conversion Table*

To convert from	To	Multiply by	
square feet	square meters	9.290304	E - 02
foot per second squared	meter per second squared	3.048	E - 01
cubic feet	cubic meters	2.831685	E - 02
pound per cubic inch	kilogram per cubic meter	2.767990	E + 04
gallon per minute	liters per second	6.309	E - 02
pound per square inch	kilopascal	6.894757	
pound force	Newton	4.448222	
kip per square foot	Pascal	4.788026	E + 04
acre-foot per day	cubic meter per second	1.427641	E - 02
acre	square meter	4.046873	E + 03
cubic foot per second	cubic meter per second	2.831685	E - 02

Note: The E indicates an exponent, as in scientific notation, followed by a positive or negative number, representing the power of 10 by which the given conversion factor is to be multiplied before use. Thus, for the square feet conversion factor, $9.290304 \times 1/100 = 0.09290304$, the factor to be used to convert square feet to square meters. For a positive exponent, as in converting acres to square meters, multiply by $4.046873 \times 1000 = 4046.8$.

Where a conversion factor cannot be found, simply use the dimensional substitution. Thus, to convert pounds per cubic inch to kilograms per cubic meter, find 1 lb = 0.4535924 kg, and 1 in^3 = 0.00001638706 m^3. Then, 1 lb/in^3 = 0.4535924 kg/0.00001638706 m^3 27,680.01, or $2.768 \times E + 4$.

*This table contains only selected values. See the U.S. Department of the Interior *Metric Manual*, or National Bureau of Standards, *The International System of Units* (SI), both available from the U.S. Government Printing Office (GPO), for far more comprehensive listings of conversion factors.

Units Used. In preparing the calculation procedures in this handbook, the editors and contributors used standard SI units throughout. In a few cases, however, certain units are still in a state of development. For example, the unit *tonne* is used in certain industries, such as waste treatment. This unit is therefore used in the waste treatment section of this handbook because it represents current practice. However, only a few SI units are still under development. Hence, users of this handbook face little difficulty from this situation.

Computer-aided Calculations. Widespread availability of programmable pocket calculators and low-cost laptop computers allows engineers and designers to save thousands of hours of calculation time. Yet each calculation procedure must be programmed, unless the engineer is willing to use off-the-shelf software. The editor—observing thousands of engineers over the years—detects reluctance among technical personnel to use untested and unproven software programs in their daily calculations. Hence, the tested and proven procedures in this handbook form excellent programming input for programmable pocket calculators, laptop computers, minicomputers, and mainframes.

A variety of software application programs can be used to put the procedures in this handbook on a computer. Typical of these are MathSoft, Algor, and similar programs.

There are a number of advantages for the engineer who programs his or her own calculation procedures, namely: (1) The engineer knows, understands, and approves *every* step in the procedure; (2) there are *no* questionable, unknown, or legally worrisome steps in the procedure; (3) the engineer has complete faith in the result because he or she knows every component of it; and (4) if a variation of the procedure is desired, it is relatively easy for the engineer to make the needed changes in the program, using this handbook as the source of the steps and equations to apply.

Modern computer equipment provides greater speed and accuracy for almost all complex design calculations. The editor hopes that engineers throughout the world will make greater use of available computing equipment in solving applied engineering problems. Becoming computer literate is a necessity for every engineer, no matter which field he or she chooses as a specialty. The procedures in this handbook simplify every engineer's task of becoming computer literate because the steps given comprise—to a great extent—the steps in the computer program that can be written.

SECTION 2
REINFORCED AND PRESTRESSED CONCRETE ENGINEERING AND DESIGN

MAX KURTZ, P.E.
Consulting Engineer

TYLER G. HICKS, P.E.
International Engineering Associates

Part 1: Reinforced Concrete

DESIGN OF FLEXURAL MEMBERS BY ULTIMATE-STRENGTH METHOD	2.4
Capacity of a Rectangular Beam	2.6
Design of a Rectangular Beam	2.6
Design of the Reinforcement in a Rectangular Beam of Given Size	2.7
Capacity of a T Beam	2.7
Capacity of a T Beam of Given Size	2.8
Design of Reinforcement in a T Beam of Given Size	2.9
Reinforcement Area for a Doubly Reinforced Rectangular Beam	2.9
Design of Web Reinforcement	2.11
Determination of Bond Stress	2.13
Design of Interior Span of a One-Way Slab	2.14
Analysis of a Two-Way Slab by the Yield-Line Theory	2.16
DESIGN OF FLEXURAL MEMBERS BY THE WORKING-STRESS METHOD	2.18
Stresses in a Rectangular Beam	2.20
Capacity of a Rectangular Beam	2.22
Design of Reinforcement in a Rectangular Beam of Given Size	2.23
Design of a Rectangular Beam	2.24
Design of Web Reinforcement	2.24
Capacity of a T Beam	2.26
Design of a T Beam Having Concrete Stressed to Capacity	2.27
Reinforcement For Doubly Reinforced Rectangular Beam	2.28
Deflection of a Continuous Beam	2.30

DESIGN OF COMPRESSION MEMBERS BY ULTIMATE-STRENGTH METHOD	2.31
Analysis of a Rectangular Member by Interaction Diagram	2.32
Axial-Load Capacity of Rectangular Member	2.33
DESIGN OF COMPRESSION MEMBERS BY WORKING-STRESS METHOD	2.35
Design of a Spirally Reinforced Column	2.35
Analysis of a Rectangular Member by Interaction Diagram	2.36
Axial-Load Capacity of a Rectangular Member	2.39
DESIGN OF COLUMN FOOTINGS	2.39
Design of an Isolated Square Footing	2.40
Combined Footing Design	2.42
CANTILEVER RETAINING WALLS	2.45
Design of a Cantilever Retaining Wall	2.46

Part 2: Prestressed Concrete

Determination of Prestress Shear and Moment	2.51
Stresses in a Beam with Straight Tendons	2.52
Determination of Capacity and Prestressing Force for a Beam with Straight Tendons	2.55
Beam with Deflected Tendons	2.57
Determination of Section Moduli	2.58
Prestressed-Concrete Beam Design Guides	2.59
Kern Distances	2.60
Magnel Diagram Construction	2.60
Camber of a Beam at Transfer	2.62
Design of a Double-T Roof Beam	2.63
Design of a Posttensioned Girder	2.67
Properties of a Parabolic Arc	2.71
Alternative Methods of Analyzing a Beam with Parabolic Trajectory	2.72
Prestress Moments in a Continuous Beam	2.73
Principle of Linear Transformation	2.75
Effect of Varying Eccentricity at End Support	2.77
Design of Trajectory for a Two-Span Continuous Beam	2.77
Steel Beam Encased in Concrete	2.84
Composite Steel-and-Concrete Beam	2.86
Design of a Concrete Joist in a Ribbed Floor	2.89
Design of a Stair Slab	2.90
Maximum Available Moment in Composite Steel and Concrete Beam	2.92

PART 1
REINFORCED CONCRETE

The design of reinforced-concrete members in this handbook is executed in accordance with the specification titled *Building Code Requirements for Reinforced Concrete* of the American Concrete Institute (ACI). The ACI *Reinforced Concrete Design Handbook* contains many useful tables that expedite design work. The designer should become thoroughly familiar with this handbook and use the tables it contains whenever possible.

The spacing of steel reinforcing bars in a concrete member is subject to the restrictions imposed by the ACI *Code*. With reference to the beam and slab shown in Fig. 1, the reinforcing steel is assumed, for simplicity, to be concentrated at its centroidal axis, and the effective depth of the flexural member is taken as the distance from the extreme compression fiber to this axis. (The term *depth* hereafter refers to the *effective* rather than the overall depth of the beam.) For design purposes, it is usually assumed that the distance from the exterior surface to the center of the first row of steel bars is 2½ in. (63.5 mm) in a beam with web stirrups, 2 in. (50.8 mm) in a beam without stirrups, and 1 in. (25.4 mm) in a slab. Where two rows of steel bars are provided, it is usually assumed that the distance from the exterior surface to the centroidal axis of the reinforcement is 3½ in. (88.9 mm). The ACI *Handbook* gives the minimum beam widths needed to accommodate various combinations of bars in one row.

In a well-proportioned beam, the width-depth ratio lies between 0.5 and 0.75. The width and overall depth are usually an even number of inches.

The basic notational system pertaining to reinforced concrete beams is as follows: f_c' = ultimate compressive strength of concrete, lb/sq.in. (kPa); f_c = maximum compressive stress in concrete, lb/sq.in. (kPa); f_s = tensile stress in steel, lb/sq.in. (kPa); f_y = yield-point stress in steel, lb/sq.in. (kPa); ε_c = strain of extreme compression fiber; ε_s = strain of steel; b = beam width, in. (mm); d = beam depth, in. (mm); A_s = area of tension reinforcement, sq.in. (cm²); p = tension-reinforcement ratio, $A_s/(bd)$; q = tension-reinforcement index, pf_y/f_c'; n = ratio of modulus of elasticity of steel to that of concrete, E_s/E_c; C = resultant compressive force on transverse section, lb (N); T = resultant tensile force on transverse section, lb (N).

Where the subscript b is appended to a symbol, it signifies that the given quantity is evaluated at balanced-design conditions.

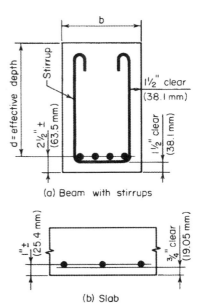

FIGURE 1. Spacing of reinforcing bars.

Design of Flexural Members by Ultimate-Strength Method

In the ultimate-strength design of a reinforced-concrete structure, as in the plastic design of a steel structure, the capacity of the structure is found by determining the load that will cause failure and dividing this result by the prescribed load factor. The load at impending failure is termed the *ultimate load*, and the maximum bending moment associated with this load is called the *ultimate moment*.

Since the tensile strength of concrete is relatively small, it is generally disregarded entirely in analyzing a beam. Consequently, the effective beam section is considered to comprise the reinforcing steel and the concrete on the compression side of the neutral axis, the concrete between these component areas serving merely as the ligature of the member.

The following notational system is applied in ultimate-strength design: a = depth of compression block, in. (mm); c = distance from extreme compression fiber to neutral axis, in. (mm); ϕ = capacity-reduction factor.

Where the subscript u is appended to a symbol, it signifies that the given quantity is evaluated at ultimate load.

For simplicity (Fig. 2), designers assume that when the ultimate moment is attained at a given section, there is a uniform stress in the concrete extending across a depth a, and that $f_c = 0.85 f_c'$, and $a = k_1 c$, where k_1 has the value stipulated in the ACI *Code*.

A reinforced-concrete beam has three potential modes of failure: crushing of the concrete, which is assumed to occur when ε_c reaches the value of 0.003; yielding of the steel, which begins when f_s reaches the value f_y; and the simultaneous crushing of the concrete and yielding of the steel. A beam that tends to fail by the third mode is said to be in *balanced design*. If the value of p exceeds that corresponding to balanced design (i.e., if there is an excess of reinforcement), the beam tends to fail by crushing of the concrete. But if the value of p is less than that corresponding to balanced design, the beam tends to fail by yielding of the steel.

Failure of the beam by the first mode would occur precipitously and without warning, whereas failure by the second mode would occur gradually, offering visible evidence of progressive failure. Therefore, to ensure that yielding of the steel would occur prior to failure of the concrete, the ACI *Code* imposes an upper limit of $0.75 p_b$ on p.

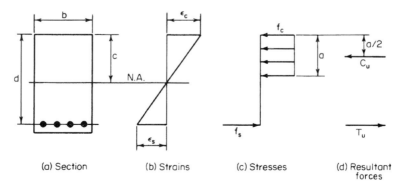

(a) Section   (b) Strains   (c) Stresses   (d) Resultant forces

FIGURE 2. Conditions at ultimate moment.

REINFORCED CONCRETE

To allow for material imperfections, defects in workmanship, etc., the *Code* introduces the capacity-reduction factor ϕ. A section of the *Code* sets $\phi = 0.90$ with respect to flexure and $\phi = 0.85$ with respect to diagonal tension, bond, and anchorage.

The basic equations for the ultimate-strength design of a rectangular beam reinforced solely in tension are

$$C_u = 0.85 abf'_c \qquad T_u = A_s f_y \tag{1}$$

$$q = \frac{[A_s/(bd)]f_y}{f'_c} \tag{2}$$

$$a = 1.18qd \qquad c = \frac{1.18qd}{k_1} \tag{3}$$

$$M_u = \phi A_s f_y \left(d - \frac{a}{2}\right) \tag{4}$$

$$M_u = \phi A_s f_y d(1 - 0.59q) \tag{5}$$

$$M_u = \phi b d^2 f'_c q(1 - 0.59q) \tag{6}$$

$$A_s = \frac{bdf_c - [(bdf_c)^2 - 2bf_c M_u/\phi]^{0.5}}{f_y} \tag{7}$$

$$p_b = \frac{0.85 k_1 f'_c}{f_y}\left(\frac{87{,}000}{87{,}000 + f_y}\right) \tag{8}$$

$$q_b = 0.85 k_1 \left(\frac{87{,}000}{87{,}000 + f_y}\right) \tag{9}$$

In accordance with the *Code*,

$$q_{\max} = 0.75 q_b = 0.6375 k_1 \left(\frac{87{,}000}{87{,}000 + f_y}\right) \tag{10}$$

Figure 3 shows the relationship between M_u and A_s for a beam of given size. As A_s increases, the internal forces C_u and T_u increase proportionately, but M_u increases by a smaller proportion because the action line of C_u is depressed. The M_u-A_s diagram is parabolic, but its curvature is small. By comparing the coordinates of two points P_a and P_b, the following result is obtained, in which the subscripts correspond to that of the given point:

$$\frac{M_{ua}}{A_{sa}} > \frac{M_{ub}}{A_{sb}} \tag{11}$$

where $A_{sa} < A_{sb}$

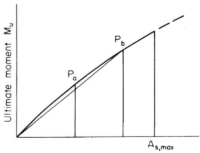

FIGURE 3

CAPACITY OF A RECTANGULAR BEAM

A rectangular beam having a width of 12 in. (304.8 mm) and an effective depth of 19.5 in. (495.3 mm) is reinforced with steel bars having an area of 5.37 sq.in. (34.647 cm²). The beam is made of 2500-lb/sq.in. (17,237.5-kPa) concrete, and the steel has a yield-point stress of 40,000 lb/sq.in. (275,800 kPa). Compute the ultimate moment this beam may resist (*a*) without referring to any design tables and without applying the basic equations of ultimate-strength design except those that are readily apparent; (*b*) by applying the basic equations.

Calculation Procedure:

1. Compute the area of reinforcement for balanced design
Use the relation $\varepsilon_s = f_y/E_s = 40,000/29,000,000 = 0.00138$. For balanced design, $c/d = \varepsilon_c/(\varepsilon_c + \varepsilon_s) = 0.003/(0.003 + 0.00138) = 0.685$. Solving for c by using the relation for c/d, we find $c = 13.36$ in. (339.344 mm). Also, $a = k_1 c = 0.85(13.36) = 11.36$ in. (288.544 mm). Then $T_u = C_u = ab(0.85)f'_c = 11.36(12)(0.85)(2500) = 290,000$ lb (1,289,920 N); $A_s = T_u/f_y = 290,000/40,000 = 7.25$ sq.in. (46,777 cm²); and $0.75A_s = 5.44$ sq.in. (35.097 cm²). In the present instance, $A_s = 5.37$ sq.in. (34.647 cm²). This is acceptable.

2. Compute the ultimate-moment capacity of this member
Thus $T_u = A_s f_y = 5.37(40,000) = 215,000$ lb (956,320 N); $C_u = ab(0.85)f'_c = 25,500a = 215,000$ lb (956,320 N); $a = 8.43$ in. (214.122 mm); $M_u = \phi T_u(d - a/2) = 0.90(215,000)(19.5 - 8.43/2) = 2,960,000$ in.·lb (334,421 N·m). These two steps comprise the solution to part *a*. The next two steps comprise the solution of part *b*.

3. Apply Eq. 10; ascertain whether the member satisfies the Code
Thus, $q_{max} = 0.6375k_1(87,000)/(87,000 + f_y) = 0.6375(0.85)(87/127) = 0.371$; $q = [A_s/(bd)]f'_c = [5.37/(12 \times 19.5)]40/2.5 = 0.367$. This is acceptable.

4. Compute the ultimate-moment capacity
Applying Eq. 5 yields $M_u = \phi A_s f_y d(1 - 0.59q) = 0.90(5.37)(40,000)(19.5)(1 - 0.59 \times 0.367) = 2,960,000$ in.·lb (334,421 N·m). This agrees exactly with the result computed in step 2.

DESIGN OF A RECTANGULAR BEAM

A beam on a simple span of 20 ft (6.1 m) is to carry a uniformly distributed live load of 1670 lb/lin ft (24,372 N/m) and a dead load of 470 lb/lin ft (6859 N/m), which includes the estimated weight of the beam. Architectural details restrict the beam width to 12 in. (304.8 mm) and require that the depth be made as small as possible. Design the section, using $f'_c = 3000$ lb/sq.in. (20,685 kPa) and $f_y = 40,000$ lb/sq.in. (275,800 kPa).

Calculation Procedure:

1. Compute the ultimate load for which the member is to be designed
The beam depth is minimized by providing the maximum amount of reinforcement permitted by the *Code*. From the previous calculation procedure, $q_{max} = 0.371$.
 Use the load factors given in the *Code*: $w_{DL} = 470$ lb/lin ft (6859 N/m); $w_{LL} = 1670$ lb/lin ft (24,372 N/m); $L = 20$ ft (6.1 m). Then $w_u = 1.5(470) + 1.8(1670) = 3710$ lb/lin ft (54,143 N/m); $M_u = \frac{1}{8}(3710)(20)^2 12 = 2,230,000$ in.·lb (251,945.4 N·m).

2. Establish the beam size
Solve Eq. 6 for d. Thus, $d^2 = M_u/[\phi b f'_c q(1 - 0.59q)] = 2,230,000/[0.90(12)(3000) \times (0.371)(0.781)]$; $d = 15.4$ in. (391.16 mm).

Set $d = 15.5$ in. (393.70 mm). Then the corresponding reduction in the value of q is negligible.

3. Select the reinforcing bars
Using Eq. 2, we find $A_s = qbdf'_c/f_y = 0.371(12)(15.5)(3/40) = 5.18$ sq.in. (33.421 cm^2). Use four no. 9 and two no. 7 bars, for which $A_s = 5.20$ sq.in. (33.550 cm^2). This group of bars cannot be accommodated in the 12-in. (304.8-mm) width and must therefore be placed in two rows. The overall beam depth will therefore be 19 in. (482.6 mm).

4. Summarize the design
Thus, the beam size is 12×19 in. (304.8×482.6 mm); reinforcement, four no. 9 and two no. 7 bars.

DESIGN OF THE REINFORCEMENT IN A RECTANGULAR BEAM OF GIVEN SIZE

A rectangular beam 9 in. (228.6 mm) wide with a 13.5-in. (342.9-mm) effective depth is to sustain an ultimate moment of 95 ft·kips (128.8 kN·m). Compute the area of reinforcement, using $f'_c = 3000$ lb/sq.in. (20,685 kPa) and $f_y = 40,000$ lb/sq.in. (275,800 kPa).

Calculation Procedure:

1. Investigate the adequacy of the beam size
From previous calculation procedures, $q_{max} = 0.371$. By Eq. 6, $M_{u,max} = 0.90 \times (9)(13.5)^2(3)(0.371)(0.781) = 1280$ in·kips (144.6 kN·m); $M_u = 95(12) = 1140$ in.·kips (128.8 kN·m). This is acceptable.

2. Apply Eq. 7 to evaluate A_s
Thus, $f_c = 0.85(3) = 2.55$ kips/sq.in. (17.582 MPa); $bdf_c = 9(13.5)(2.55) = 309.8$ kips (1377.99 kN); $A_s = [309.8 - (309.8^2 - 58,140)^{0.5}]/40 = 2.88$ sq.in. (18.582 cm^2).

CAPACITY OF A T BEAM

Determine the ultimate moment that may be resisted by the T beam in Fig. 4a if $f'_c = 3000$ lb/sq.in. (20,685 kPa) and $f_y = 40,000$ lb/sq.in. (275,800 kPa).

Calculation Procedure:

1. Compute T_u and the resultant force that may be developed in the flange
Thus, $T_u = 8.20(40,000) = 328,000$ lb (1,458,944 N); $f_c = 0.85(3000) = 2550$ lb/sq.in. (17,582.3 kPa); $C_{uf} = 18(6)(2550) = 275,400$ lb (1,224,979 N). Since $C_{uf} < T_u$, the deficiency must be supplied by the web.

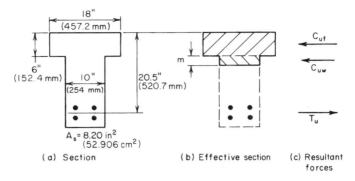

(a) Section   (b) Effective section   (c) Resultant forces

FIGURE 4

2. Compute the resultant force developed in the web and the depth of the stress block in the web

Thus, $C_{uw} = 328{,}000 - 275{,}400 = 52{,}600$ lb (233,964.8 N); m = depth of the stress block = $52{,}600/[2550(10)] = 2.06$ in. (52.324 mm).

3. Evaluate the ultimate-moment capacity

Thus, $M_u = 0.90[275{,}400(20.5 - 3) + 52{,}600(20.5 - 6 - 1.03)] = 4{,}975{,}000$ in.·lb (562,075.5 N·m).

4. Determine if the reinforcement complies with the Code

Let b' = width of web, in. (mm); A_{s1} = area of reinforcement needed to resist the compressive force in the overhanging portion of the flange, sq.in. (cm²); A_{s2} = area of reinforcement needed to resist the compressive force in the remainder of the section, sq.in. (cm²). Then $p_2 = A_{s2}/(b'd)$; $A_{s1} = 2550(6)(18 - 10)740{,}000 = 3.06$ sq.in. (19.743 cm²); $A_{s2} = 8.20 - 3.06 = 5.14$ sq.in. (33.163 cm²). Then $p_2 = 5.14/[10(20.5)] = 0.025$.

A section of the ACI *Code* subjects the reinforcement ratio p_2 to the same restriction as that in a rectangular beam. By Eq. 8, $p_{2,\max} = 0.75 p_b = 0.75(0.85)(0.85)(3/40)(87/127) = 0.0278 > 0.025$. This is acceptable.

CAPACITY OF A T BEAM OF GIVEN SIZE

The T beam in Fig. 5 is made of 3000-lb/sq.in. (20,685-kPa) concrete, and $f_y = 40{,}000$ lb/sq.in. (275,800 kPa). Determine the ultimate-moment capacity of this member if it is reinforced in tension only.

Calculation Procedure:

1. Compute C_{u1}, $C_{u2,\max}$, and $s_{\max}$

Let the subscript 1 refer to the overhanging portion of the flange and the subscript 2 refer to the remainder of the compression zone. Then $f_c = 0.85(3000) = 2550$ lb/sq.in. (17,582.3 kPa); $C_{u1} = 2550(5)(16 - 10) = 76{,}500$ lb (340,272 N). From the previous calculation procedure, $p_{2,\max} = 0.0278$. Then $A_{s2,\max} = 0.0278(10)(19.5) = 5.42$ sq.in. (34.970 cm²); $C_{u2,\max} =$

5.42(40,000) = 216,800 lb (964,326.4 N); s_{max} = 216,800/[10(2550)] = 8.50 in. (215.9 mm).

2. Compute the ultimate-moment capacity
Thus, $M_{u,max}$ = 0.90[(76,500(19.5 − 5/2) + 216,800(19.5 − 8.50/2)] = 4,145,000 in.·lb (468,300 N·m).

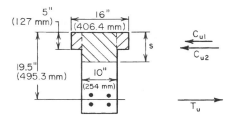

FIGURE 5

DESIGN OF REINFORCEMENT IN A T BEAM OF GIVEN SIZE

The T beam in Fig. 5 is to resist an ultimate moment of 3,960,000 in.·lb (447,400.8 N·m). Determine the required area of reinforcement, using f'_c = 3000 lb/sq.in. (20,685 kPa) and f_y = 40,000 lb/sq.in. (275,800 kPa).

Calculation Procedure:

1. Obtain a moment not subject to reduction
From the previous calculation procedure, the ultimate-moment capacity of this member is 4,145,000 in.·lb (468,300 N·m). To facilitate the design, divide the given ultimate moment M_u by the capacity-reduction factor to obtain a moment M'_u that is not subject to reduction. Thus M'_u = 3,960,000/0.9 = 4,400,000 in.·lb (497,112 N·m).

2. Compute the value of s associated with the given moment
From step 2 in the previous calculation procedure, M'_{u1} = 1,300,000 in.·lb (146,874 N·m). Then M'_{u2} = 4,400,000 − 1,300,000 = 3,100,000 in.·lb (350,238 N·m). But M'_{u2} = 2550(10s)(19.5 − s/2), so s = 7.79 in. (197.866 mm).

3. Compute the area of reinforcement
Thus, $C_{u2} = M'_{u2}/(d − ½s)$ = (19.5 − 3.90) = 198,700 lb (883,817.6 N). From step 1 of the previous calculation procedure, C_{u1} = 76,500 lb (340,272 N); T_u = 76,500 + 198,700 = 275,200 lb (1,224,089.6 N); A_s = 275,200/40,000 = 6.88 sq.in. (174.752 mm).

4. Verify the solution
To verify the solution, compute the ultimate-moment capacity of the member. Use the notational system given in earlier calculation procedures. Thus, C_{uf} = 16(5)(2550) = 204,000 lb (907,392 N); C_{uw} = 275,200 − 204,000 = 71,200 lb (316,697.6 N); m = 71,200/[2550(10)] = 2.79 in. (70.866 mm); M_u = 0.90 [204,000(19.5 − 2.5) + 71,200(19.5 − 5 − 1.40)] = 3,960,000 in.·lb (447,400.8 N·m). Thus, the result is verified because the computed moment equals the given moment.

REINFORCEMENT AREA FOR A DOUBLY REINFORCED RECTANGULAR BEAM

A beam that is to resist an ultimate moment of 690 ft·kips (935.6 kN·m) is restricted to a 14-in. (355.6-mm) width and 24-in. (609.6-mm) total depth. Using f'_c = 5000 lb/sq.in. and f_y = 50,000 lb/sq.in. (344,750 kPa), determine the area of reinforcement.

Calculation Procedure:

1. Compute the values of q_b, q_{max}, and p_{max} for a singly reinforced beam

As the following calculations will show, it is necessary to reinforce the beam both in tension and in compression. In Fig. 6, let A_s = area of tension reinforcement, sq.in. (cm^2); A'_s = area of compression reinforcement, sq.in. (cm^2); d' = distance from compression face of concrete to centroid of compression reinforcement, in. (mm); f_s = stress in tension steel, lb/sq.in. (kPa); f'_s = stress in compression steel, lb/sq.in. (kPa); ε'_s = strain in compression steel; $p = A_s/(bd)$; $p' = A'_s/(bd)$; $q = pf_y/f'_c$; M_u = ultimate moment to be resisted by member, in.·lb (N·m); M_{u1} = ultimate-moment capacity of member if reinforced solely in tension; M_{u2} = increase in ultimate-moment capacity resulting from use of compression reinforcement; C_{u1} = resultant force in concrete, lb (N); C_{u2} = resultant force in compression steel, lb (N).

If $f'_s = f_y$, the tension reinforcement may be resolved into two parts having areas of $A_s - A'_s$ and A'_s. The first part, acting in combination with the concrete, develops the moment M_{u1}. The second part, acting in combination with the compression reinforcement, develops the moment M_{s2}.

To ensure that failure will result from yielding of the tension steel rather than crushing of the concrete, the ACI *Code* limits $p - p'$ to a maximum value of $0.75p_b$, where p_b has the same significance as for a singly reinforced beam. Thus the *Code*, in effect, permits setting $f'_s = f_y$ if inception of yielding in the compression steel will precede or coincide with failure of the concrete at balanced-design ultimate moment. This, however, introduces an inconsistency, for the limit imposed on $p - p'$ precludes balanced design.

By Eq. 9, $q_b = 0.85(0.80)(87/137) = 0.432$; $q_{max} = 0.75(0.432) = 0.324$; $p_{max} = 0.324(5/50) = 0.0324$.

2. Compute M_{u1}, M_{u2}, and C_{u2}

Thus, $M_u = 690{,}000(12) = 8{,}280{,}000$ in.·lb (935,474.4 N·m). Since two rows of tension bars are probably required, $d = 24 - 3.5 = 20.5$ in. (520.7 mm). By Eq. 6, $M_{u1} = 0.90(14)(20.5)^2(5000) \times (0.324)(0.809) = 6{,}940{,}000$ in.·lb (784,081.2 N·m); $M_{u2} = 8{,}280{,}000 - 6{,}940{,}000 = 1{,}340{,}000$ in.·lb (151,393.2 N·m); $C_{u2} = M_{u2}/(d - d') = 1{,}340{,}000/(20.5 - 2.5) = 74{,}400$ lb (330,931.2 N).

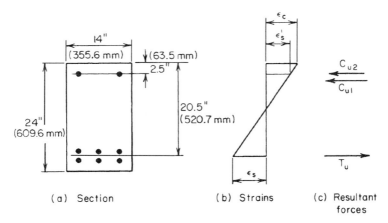

(a) Section (b) Strains (c) Resultant forces

FIGURE 6. Doubly reinforced rectangular beam.

3. Compute the value of ε'_s under the balanced-design ultimate moment
Compare this value with the strain at incipient yielding. By Eq. 3, $c_b = 1.18 q_b d/k_1 = 1.18(0.432)(20.5)/0.80 = 13.1$ in. (332.74 mm); $\varepsilon'_s/\varepsilon_c = (13.1 - 2.5)/13.1 = 0.809$; $\varepsilon'_s = 0.809(0.003) = 0.00243$; $\varepsilon_y = 50/29{,}000 = 0.0017 < \varepsilon'_s$. The compression reinforcement will therefore yield before the concrete fails, and $f'_s = f_y$ may be used.

4. Alternatively, test the compression steel for yielding
Apply

$$p - p' \geq \frac{0.85 k_1 f'_c d'(87{,}000)}{f_y d (87{,}000 - f_y)} \tag{12}$$

If this relation obtains, the compression steel will yield. The value of the right-hand member is $0.85(0.80)(5/50)(2.5/20.5)(87/37) = 0.0195$. From the preceding calculations, $p - p' = 0.0324 > 0.0195$. This is acceptable.

5. Determine the areas of reinforcement
By Eq. 2, $A_s = A'_s = q_{max} b d f'_c / f_y = 0.324(14)(20.5)(5/50) = 9.30$ sq.in. (60.00 cm²); $A'_s = C_{u2}/(\phi f_y) = 74{,}400/[0.90(50{,}000)] = 1.65$ sq.in. (10.646 cm²); $A_s = 9.30 + 1.65 = 10.95$ sq.in. (70.649 cm²).

6. Verify the solution
Apply the following equations for the ultimate-moment capacity:

$$a = \frac{(A_s - A'_s) f_y}{0.85 f'_c b} \tag{13}$$

So $a = 9.30(50{,}000)/[0.85(5000)(14)] = 7.82$ in. (198.628 mm). Also,

$$M_u = \phi f_y \left[(A_s - A'_s)\left(d - \frac{a}{2}\right) + A'_s(d - d') \right] \tag{14}$$

So $M_u = 0.90(50{,}000)(9.30 \times (16.59 + 1.65 \times 18) = 8{,}280{,}000$ in.·lb (935,474.4 N·m), as before. Therefore, the solution has been verified.

DESIGN OF WEB REINFORCEMENT

A 15-in. (381-mm) wide 22.5-in. (571.5-mm) effective-depth beam carries a uniform ultimate load of 10.2 kips/lin ft (148.86 kN/m). The beam is simply supported, and the clear distance between supports is 18 ft (5.5 m). Using $f'_c = 3000$ lb/sq.in. (20,685 kPa) and $f_y = 40{,}000$ lb/sq.in. (275,800 kPa), design web reinforcement in the form of vertical U stirrups for this beam.

Calculation Procedure:

1. Construct the shearing-stress diagram for half-span
The ACI *Code* provides two alternative methods for computing the allowable shearing stress on an unreinforced web. The more precise method recognizes the contribution of both the shearing stress and flexural stress on a cross section in producing diagonal

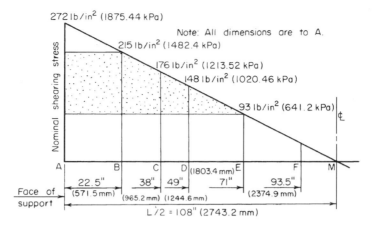

FIGURE 7. Shearing stress diagram.

tension. The less precise and more conservative method restricts the shearing stress to a stipulated value that is independent of the flexural stress.

For simplicity, the latter method is adopted here. A section of the *Code* sets $\phi = 0.85$ with respect to the design of web reinforcement. Let v_u = nominal ultimate shearing stress, lb/sq.in. (kPa); v_c = shearing stress resisted by concrete, lb/sq.in. (kPa); v_u' = shearing stress resisted by the web reinforcement, lb/sq.in. (kPa); A_v = total cross-sectional area of stirrup, sq.in. (cm^2); V_u = ultimate vertical shear at section, lb (N); s = center-to-center spacing of stirrups, in. (mm).

The shearing-stress diagram for half-span is shown in Fig. 7. Establish the region *AF* within which web reinforcement is required. The *Code* sets the allowable shearing stress in the concrete at

$$v_c = 2\phi(f_c')^{0.5} \tag{15}$$

The equation for nominal ultimate shearing stress is

$$v_u = \frac{V_u}{bd} \tag{16}$$

Then, $v_c = 2(0.85)(3000)^{0.5} = 93$ lb/sq.in. (641.2 kPa).

At the face of the support, $V_u = 9(10,200) = 91,800$ lb (408,326.4 N); $v_u = 91,800/[15(22.5)] = 272$ lb/sq.in. (1875.44 kPa). The slope of the shearing-stress diagram = $-272/108 = -2.52$ lb/(in^2·in.) (−0.684 kPa/mm). At distance d from the face of the support, $v_u = 272 - 22.5(2.52) = 215$ lb/sq.in. (1482.4 kPa); $v_u' = 215 - 93 = 122$ lb/sq.in. (841.2 kPa).

Let *E* denote the section at which $v_u = v_c$. Then, $AE = (272 - 93)/2.52 = 71$ in. (1803.4 mm). A section of the *Code* requires that web reinforcement be continued for a distance d beyond the section where $v_u = v_c$; $AF = 71 + 22.5 = 93.5$ in. (2374.9 mm).

2. Check the beam size for Code compliance

Thus, $v_{u,\max} = 10\phi(f_c')^{0.5} = 466 > 215$ lb/sq.in. (1482.4 kPa). This is acceptable.

3. Select the stirrup size
Equate the spacing near the support to the minimum practical value, which is generally considered to be 4 in. (101.6 mm). The equation for stirrup spacing is

$$s = \frac{\phi A_v f_y}{v'_c b} \qquad (17)$$

Then $A_v = s v'_u b/(\phi f_y) = 4(122)(15)/[0.85(40,000)] = 0.215$ sq.in. (1.3871 cm²). Since each stirrup is bent into the form of a U, the total cross-sectional area is twice that of a straight bar. Use no. 3 stirrups for which $A_v = 2(0.11) = 0.22$ sq.in. (1.419 cm²).

4. Establish the maximum allowable stirrup spacing
Apply the criteria of the *Code*, or $s_{max} = d/4$ if $v > 6\phi(f'_c)^{0.5}$. The right-hand member of this inequality has the value 279 lb/sq.in. (1923.70 kPa), and this limit therefore does not apply. Then $s_{max} = d/2 = 11.25$ in. (285.75 mm), or $s_{max} = A_v/(0.0015b) = 0.22/[0.0015(15)] = 9.8$ in. (248.92 mm). The latter limit applies, and the stirrup spacing will therefore be restricted to 9 in. (228.6 mm).

5. Locate the beam sections at which the required stirrup spacing is 6 in. (152.4 mm) and 9 in. (228.6 mm)
Use Eq. 17. Then $\phi A_v f_y/b = 0.85(0.22)(40,000)/15 = 499$ lb/in. (87.38 kN/m). At C: $v'_u = 499/6 = 83$ lb/sq.in. (572.3 kPa); $v_u = 83 + 93 = 176$ lb/sq.in. (1213.52 kPa); $AC = (272 - 176)/2.52 = 38$ in. (965.2 mm). At D: $v'_u = 499/9 = 55$ lb/sq.in. (379.2 kPa); $v_u = 55 + 93 = 148$ lb/sq.in. (1020.46 kPa); $AD = (272 - 148)/2.52 = 49$ in. (1244.6 mm).

6. Devise a stirrup spacing conforming to the computed results
The following spacing, which requires 17 stirrups for each half of the span, is satisfactory and conforms with the foregoing results:

Quantity	Spacing, in. (mm)	Total, in. (mm)	Distance from last stirrup to face of support, in. (mm)
1	2 (50.8)	2 (50.8)	2 (50.8)
9	4 (101.6)	36 (914.4)	38 (965.2)
2	6 (152.4)	12 (304.8)	50 (1270)
5	9 (228.6)	45 (1143)	95 (2413)

DETERMINATION OF BOND STRESS

A beam of 4000-lb/sq.in. (27,580-kPa) concrete has an effective depth of 15 in. (381 mm) and is reinforced with four no. 7 bars. Determine the ultimate bond stress at a section where the ultimate shear is 72 kips (320.3 kN). Compare this with the allowable stress.

Calculation Procedure:

1. Determine the ultimate shear flow h_u
The adhesion of the concrete and steel must be sufficiently strong to resist the horizontal shear flow. Let u_u = ultimate bond stress, lb/sq.in. (kPa); V_u = ultimate vertical shear, lb (N);

Σo = sum of perimeters of reinforcing bars, in. (mm). Then the ultimate shear flow at any plane between the neutral axis and the reinforcing steel is $h_u = V_u/(d - a/2)$.

In conformity with the notational system of the working-stress method, the distance $d - a/2$ is designated as jd. Dividing the shear flow by the area of contact in a unit length and introducing the capacity-reduction factor yield

$$u_u = \frac{V_u}{\phi \Sigma o \, jd} \tag{18}$$

A section of the ACI *Code* sets $\phi = 0.85$ with respect to bond, and j is usually assigned the approximate value of 0.875 when this equation is used.

2. Calculate the bond stress
Thus, $\Sigma o = 11.0$ in. (279.4 mm), from the ACI *Handbook*. Then $u_u = 72,000/[0.85(11.0)(0.875) \times (15)] = 587$ lb/sq.in. (4047.4 kPa).

The allowable stress is given in the *Code* as

$$u_{u,\text{allow}} = \frac{9.5(f'_c)^{0.5}}{D} \tag{19}$$

but not above 800 lb/sq.in. (5516 kPa). Thus, $u_{u,\text{allow}} = 9.5(4,000)^{0.5}/0.875 = 687$ lb/sq.in. (4736.9 kPa).

DESIGN OF INTERIOR SPAN OF A ONE-WAY SLAB

A floor slab that is continuous over several spans carries a live load of 120 lb/sq.ft. (5745 N/m²) and a dead load of 40 lb/sq.ft. (1915 N/m²), exclusive of its own weight. The clear spans are 16 ft (4.9 m). Design the interior span, using $f'_c = 3000$ lb/sq.in. (20,685 kPa) and $f_y = 50,000$ lb/sq.in. (344,750 kPa).

Calculation Procedure:

1. Find the minimum thickness of the slab as governed by the Code
Refer to Fig. 8. The maximum potential positive or negative moment may be found by applying the type of loading that will induce the critical moment and then evaluating this moment. However, such an analysis is time-consuming. Hence, it is wise to apply the moment equations recommended in the ACI *Code* whenever the span and loading conditions satisfy the requirements given there. The slab is designed by considering a 12-in. (304.8-mm) strip as an individual beam, making $b = 12$ in. (304.8 mm).

Assuming that $L = 17$ ft (5.2 m), we know the minimum thickness of the slab is $t_{\min} = L/35 = 17(12)/35 = 5.8$ in. (147.32 mm).

2. Assuming a slab thickness, compute the ultimate load on the member
Tentatively assume $t = 6$ in. (152.4 mm). Then the beam weight = $(6/12)(150$ lb/ft³$) = 75$ lb/lin ft (1094.5 N/m). Also, $w_u = 1.5(40 + 75) + 1.8(120) = 390$ lb/lin ft (5691.6 N/m).

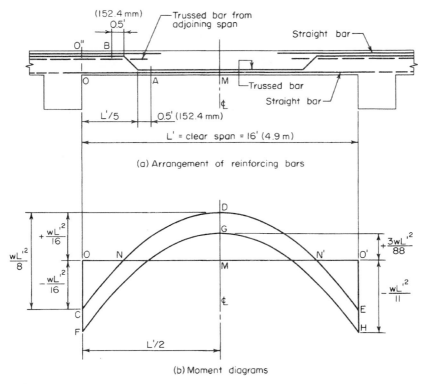

FIGURE 8

3. Compute the shearing stress associated with the assumed beam size

From the *Code* for an interior span, $V_u = \frac{1}{2}w_uL' = \frac{1}{2}(390)(16) = 3120$ lb (13,877.8 N); $d = 6 - 1 = 5$ in. (127 mm); $v_u = 3120/[12(5)] = 52$ lb/sq.in. (358.54 kPa); $v_c = 93$ lb/sq.in. (641.2 kPa). This is acceptable.

4. Compute the two critical moments

Apply the appropriate moment equations. Compare the computed moments with the moment capacity of the assumed beam size to ascertain whether the size is adequate. Thus, $M_{u,\text{neg}} = (^1/_{11})w_uL'^2 = (^1/_{11})(390)(16)^2(12) = 108{,}900$ in.·lb (12,305.5 N·m), where the value 12 converts the dimension to inches. Then $M_{u,\text{pos}} = ^1/_{16}w_uL'^2 = 74{,}900$ in.·lb (8462.2 N·m). By Eq. 10, $q_{\max} = 0.6375(0.85)(87/137) = 0.344$. By Eq. 6, $M_{u,\text{allow}} = 0.90(12)(5)^2(3000)(0.344)(0.797) = 222{,}000$ in.·lb (25,081.5 N·m). This is acceptable. The slab thickness will therefore be made 6 in. (152.4 mm).

5. Compute the area of reinforcement associated with each critical moment

By Eq. 7, $bdf_c = 12(5)(2.55) = 153.0$ kips (680.54 kN); then $2bf_cM_{u,\text{neg}}/\phi = 2(12)(2.55)(108.9)/0.90 = 7405$ kips^2 (146,505.7 kN^2); $A_{s,\text{neg}} = [153.0 - (153.0^2 - 7405)^{0.5}]/50 = 0.530$ sq.in. (3.4196 cm^2). Similarly, $A_{s,\text{pos}} = 0.353$ sq.in. (2.278 cm^2).

6. Select the reinforcing bars, and locate the bend points

For positive reinforcement, use no. 4 trussed bars 13 in. (330.2 mm) on centers, alternating with no. 4 straight bars 13 in. (330.2 mm) on centers, thus obtaining $A_s = 0.362$ sq.in. (2.336 cm^2).

For negative reinforcement, supplement the trussed bars over the support with no. 4 straight bars 13 in. (330.2 mm) on centers, thus obtaining $A_s = 0.543$ sq.in. (3.502 cm^2).

The trussed bars are usually bent upward at the fifth points, as shown in Fig. 8a. The reinforcement satisfies a section of the ACI *Code* which requires that "at least ... one-fourth the positive moment reinforcement in continuous beams shall extend along the same face of the beam into the support at least 6 in. (152.4 mm)."

7. Investigate the adequacy of the reinforcement beyond the bend points

In accordance with the *Code*, $A_{min} = A_t = 0.0020bt = 0.0020(12)(6) = 0.144$ sq.in. (0.929 cm^2).

A section of the *Code* requires that reinforcing bars be extended beyond the point at which they become superfluous with respect to flexure a distance equal to the effective depth or 12 bar diameters, whichever is greater. In the present instance, extension = 12(0.5) = 6 in. (152.4 mm). Therefore, the trussed bars in effect terminate as positive reinforcement at section A (Fig. 8). Then $L'/5 = 3.2$ ft (0.98 m); $AM = 8 - 3.2 - 0.5 = 4.3$ ft (1.31 m).

The conditions immediately to the left of A are $M_u = M_{u,pos} - \frac{1}{2}w_u(AM)^2 = 74,900 - \frac{1}{2}(390)(4.3)^2(12) = 31,630$ in.·lb (3573.56 N·m); $A_{s,pos} = 0.181$ sq.in. (1.168 cm^2); $q = 0.181(50)/[12(5)(3)] = 0.0503$. By Eq. 5, $M_{u,allow} = 0.90(0.181)(50,000)(5)(0.970) = 39,500$ in.·lb (4462.7 N·m). This is acceptable.

Alternatively, Eq. 11 may be applied to obtain the following conservative approximation: $M_{u,allow} = 74,900(0.181)/0.353 = 38,400$ in.·lb (4338.43 N·m).

The trussed bars in effect terminate as negative reinforcement at B, where $O''B = 3.2 - 0.33 - 0.5 = 2.37$ ft (72.23 m). The conditions immediately to the right of B are $|M_u| = M_{u,neg} - 12(3120 \times 2.37 - \frac{1}{2} \times 390 \times 2.37^2) = 33,300$ in.·lb (3762.23 N·m). Then $A_{s,neg} = 0.362$ sq.in. (2.336 cm^2). As a conservative approximation, $M_{u,allow} = 108,900(0.362)/0.530 = 74,400$ in.·lb (8405.71 N·m). This is acceptable.

8. Locate the point at which the straight bars at the top may be discontinued

9. Investigate the bond stresses

In accordance with Eq. 19, $U_{u,allow} = 800$ lb/sq.in. (5516 kPa).

If CDE in Fig. 8b represents the true moment diagram, the bottom bars are subjected to bending stress in the interval NN'. Manifestly, the maximum bond stress along the bottom occurs at these boundary points (points of contraflexure), where the shear is relatively high and the straight bars alone are present. Thus $MN = 0.354L'$; V_u at N/V_u at support = $0.354L'/(0.5L') = 0.71$; V_u at $N = 0.71(3120) = 2215$ lb (9852.3 N). By Eq. 18, $u_u = V_u/(\phi \Sigma o j d) = 2215/[0.85(1.45)(0.875)(5)] = 411$ lb/sq.in. (2833.8 kPa). This is acceptable. It is apparent that the maximum bond stress in the top bars has a smaller value.

ANALYSIS OF A TWO-WAY SLAB BY THE YIELD-LINE THEORY

The slab in Fig. 9a is simply supported along all four edges and is isotropically reinforced. It supports a uniformly distributed ultimate load of w_u lb/sq.ft. (kPa). Calculate the ultimate unit moment m_u for which the slab must be designed.

REINFORCED CONCRETE

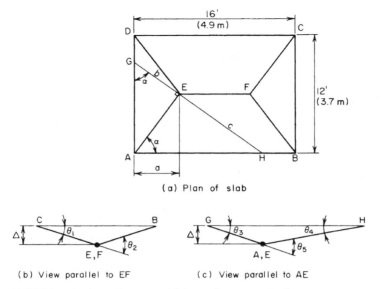

FIGURE 9. Analysis of two-way slab by mechanism method.

Calculation Procedure:

1. Draw line GH perpendicular to AE at E; express distances b and c in terms of a

Consider a slab to be reinforced in orthogonal directions. If the reinforcement in one direction is identical with that in the other direction, the slab is said to be *isotropically reinforced*; if the reinforcements differ, the slab is described as *orthogonally anisotropic*. In the former case, the capacity of the slab is identical in all directions; in the latter case, the capacity has a unique value in every direction. In this instance, assume that the slab size is excessive with respect to balanced design, the result being that the failure of the slab will be characterized by yielding of the steel.

In a steel beam, a plastic hinge forms at a *section*; in a slab, a plastic hinge is assumed to form along a straight line, termed a yield line. It is plausible to assume that by virtue of symmetry of loading and support conditions the slab in Fig. 9a will fail by the formation of a central yield line *EF* and diagonal yield lines such as *AE*, the ultimate moment at these lines being positive. The ultimate *unit* moment m_u is the moment acting on a unit length.

Although it is possible to derive equations that give the location of the yield lines, this procedure is not feasible because the resulting equations would be unduly cumbersome. The procedure followed in practice is to assign a group of values to the distance *a* and to determine the corresponding values of m_u. The true value of m_u is the highest one obtained. Either the static or mechanism method of analysis may be applied; the latter will be applied here.

Expressing the distances *b* and *c* in terms of *a* gives tan $a = 6/a = AE/b = c/(AE)$; $b = aAE/6$; $c = 6AE/a$.

2. Find the rotation of the plastic hinges

Allow line *EF* to undergo a virtual displacement Δ after the collapse load is reached. During the virtual displacement, the portions of the slab bounded by the yield lines and the supports rotate as planes. Refer to Fig. 9b and c: $\theta_1 = \Delta/6$; $\theta_2 = 2\theta_1 = \Delta/3 = 0.333\Delta$; $\theta_3 = \Delta/b$; $\theta_4 = \Delta/c$; $\theta_5 = \Delta(1/b + 1/c) = [\Delta/(AE)](6/a + a/6)$.

3. Select a trial value of a, and evaluate the distances and angles
Using $a = 4.5$ ft (1.37 m) as the trial value, we find $AE = (a^2 + 6^2)^{0.5} = 7.5$ ft (2.28 m); $b = 5.63$ ft (1.716 m); $c = 10$ ft (3.0 m); $\theta_5 = (\Delta/7.5)(6/4.5 + 4.5/6) = 0.278\Delta$.

4. Develop an equation for the external work W_E performed by the uniform load on a surface that rotates about a horizontal axis
In Fig. 10, consider that the surface ABC rotates about axis AB through an angle θ while carrying a uniform load of w lb/sq.ft. (kPa). For the elemental area dA_s, the deflection, total load, and external work are $\delta = x\theta$; $dW = w\,dA$; $dW_E = \delta\,dW = x\theta w\,dA$. The total work for the surface is $W_E = w\theta \int x\,dA$, or

$$W_E = w\theta Q \qquad (20)$$

where Q = static moment of total area, with respect to the axis of rotation.

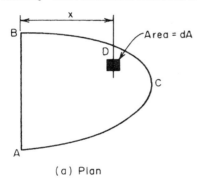

(a) Plan

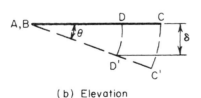

(b) Elevation

FIGURE 10

5. Evaluate the external and internal work for the slab
Using the assumed value, we see $a = 4.5$ ft (1.37 m), $EF = 16 - 9 = 7$ ft (2.1 m). The external work for the two triangles is $2w_u(\Delta/4.5)(1/6)(12)(4.5)^2 = 18w_u\Delta$. The external work for the two trapezoids is $2w_u(\Delta/6)(1/6)(16 + 2 \times 7)(6)^2 = 60w_u\Delta$. Then $W_E = w_u\Delta(18 + 60) = 78w_u\Delta$; $W_I = m_u(7\theta_2 + 4 \times 7.5\theta_5) = 10.67m_u\Delta$.

6. Find the value of m_u corresponding to the assumed value of a
Equate the external and internal work to find this value of m_u. Thus, $10.67m_u\Delta = 78w_u\Delta$; $m_u = 7.31w_u$.

7. Determine the highest value of m_u
Assign other trial values to a, and find the corresponding values of m_u. Continue this procedure until the highest value of m_u is obtained. This is the true value of the ultimate unit moment.

Design of Flexural Members by the Working-Stress Method

As demonstrated earlier, the analysis or design of a composite beam by the working-stress method is most readily performed by transforming the given beam to an equivalent homogeneous beam. In the case of a reinforced-concrete member, the transformation is made by replacing the reinforcing steel with a strip of concrete having an area nA_s and located at the same distance from the neutral axis as the steel. This substitute concrete is assumed capable of sustaining tensile stresses.

The following symbols, shown in Fig. 11, are to be added to the notational system given earlier: kd = distance from extreme compression fiber to neutral axis, in. (mm); jd = distance between action lines of C and T, in. (mm); z = distance from extreme compression fiber to action line of C, in. (mm).

The basic equations for the working-stress design of a rectangular beam reinforced solely in tension are

$$k = \frac{f_c}{f_c + f_s/n} \quad (21)$$

$$j = 1 - \frac{k}{3} \quad (22)$$

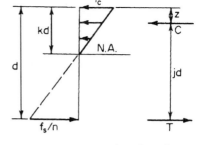

FIGURE 11. Stress and resultant forces.

$$M = cjd = \frac{1}{2} f_c kjbd^2 \quad (23)$$

$$M = \frac{1}{6} f_c k(3-k)bd^2 \quad (24)$$

$$M = Tjd = f_s A_s jd \quad (25)$$

$$M = f_s pjbd^2 \quad (26)$$

$$M = \frac{f_s k^2 (3-k) bd^2}{6n(1-k)} \quad (27)$$

$$p = \frac{f_c k}{2 f_s} \quad (28)$$

$$p = \frac{k^2}{2n(1-k)} \quad (29)$$

$$k = [2pn + (pn)^2]^{0.5} - pn \quad (30)$$

For a given set of values of f_c, f_s, and n, M is directly proportional to the beam property bd^2. Let K denote the constant of proportionality. Then

$$M = Kbd^2 \quad (31)$$

where

$$K = \frac{1}{2} f_c kj = f_s pj \quad (32)$$

The allowable flexural stress in the concrete and the value of n, which are functions of the ultimate strength f'_c, are given in the ACI *Code*, as is the allowable flexural stress in the steel. In all instances in the following procedures, the assumption is that the reinforcement is intermediate-grade steel having an allowable stress of 20,000 lb/sq.in. (137,900 kPa).

TABLE 1. Values of Design Parameters at Balanced Design

f'_c and n	f_c	f_s	K	k	j	p
2500	1125	20,000	178	0.360	0.880	0.0101
10						
3000	1350	20,000	223	0.378	0.874	0.0128
9						
4000	1800	20,000	324	0.419	0.853	0.0188
8						
5000	2250	20,000	423	0.441	0.853	0.0248
7						

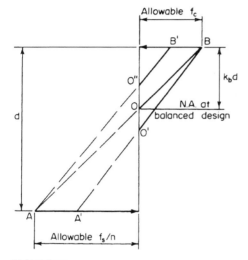

FIGURE 12

Consider that the load on a beam is gradually increased until a limiting stress is induced. A beam that is so proportioned that the steel and concrete simultaneously attain their limiting stress is said to be in *balanced design*. For each set of values of f'_c and f_s, there is a corresponding set of values of K, k, j, and p associated with balanced design. These values are recorded in Table 1.

In Fig. 12, AB represents the stress line of the transformed section for a beam in balanced design. If the area of reinforcement is increased while the width and depth remain constant, the neutral axis is depressed to O', and $A'O'B$ represents the stress line under the allowable load. But if the width is increased while the depth and area of reinforcement remain constant, the neutral axis is elevated to O'', and $AO''B'$ represents the stress line under the allowable load. This analysis leads to these conclusions: If the reinforcement is in excess of that needed for balanced design, the concrete is the first material to reach its limiting stress under a gradually increasing load. If the beam size is in excess of that needed for balanced design, the steel is the first material to reach its limiting stress.

STRESSES IN A RECTANGULAR BEAM

A beam of 2500-lb/sq.in (17,237.5-kPa) concrete has a width of 12 in. (304.8 mm) and an effective depth of 19.5 in. (495.3 mm). It is reinforced with one no. 9 and two no. 7 bars. Determine the flexural stresses caused by a bending moment of 62 ft·kips (84.1 kN·m) (*a*) without applying the basic equations of reinforced-concrete beam design; (*b*) by applying the basic equations.

Calculation Procedure:

1. Record the pertinent beam data
Thus $f'_c = 2500$ lb/sq.in. (17,237.5 kPa); $\therefore n = 10$; $A_s = 2.20$ sq.in. (14.194 cm^2); $nA_s = 22.0$ sq.in. (141.94 cm^2). Then $M = 62,000(12) = 744,000$ in.·lb (84,057.1 N·m).

2. Transform the given section to an equivalent homogeneous section, as in Fig. 13b

3. Locate the neutral axis of the member
The neutral axis coincides with the centroidal axis of the transformed section. To locate the neutral axis, set the static moment of the transformed area with respect to its centroidal axis equal to zero: $12(kd)^2/2 - 22.0(19.5 - kd) = 0$; $kd = 6.82$; $d - kd = 12.68$ in. (322.072 mm).

4. Calculate the moment of inertia of the transformed section
Then evaluate the flexural stresses by applying the stress equation: $I = (\frac{1}{3})(12)(6.82)^3 + 22.0(12.68)^2 = 4806$ in^4 (200,040.6 cm^4); $f_c = Mkd/I = 744,000(6.82)/4806 = 1060$ lb/sq.in. (7308.7 kPa); $f_s, = 10(744,000)(12.68)/4806 = 19,600$ lb/sq.in.

5. Alternatively, evaluate the stresses by computing the resultant forces C and T
Thus $jd = 19.5 - 6.82/3 = 17.23$ in. (437.642 mm); $C = T = M/jd = 744,000/17.23 = 43,200$ lb (192,153.6 N). But $C = \frac{1}{2}f_c(6.82)12$; $\therefore f_c = 1060$ lb/sq.in. (7308.7 kPa); and $T = 2.20f_s$; $\therefore f_s = 19,600$ lb/sq.in. (135,142 kPa). This concludes part a of the solution. The next step constitutes the solution to part b.

6. Compute pn and then apply the basic equations in the proper sequence
Thus $p = A_s/(bd) = 2.20/[12(19.5)] = 0.00940$; $pn = 0.0940$. Then by Eq. 30, $k = [0.188 + (0.094)^2]^{0.5} - 0.094 = 0.350$. By Eq. 22, $j = 1 - 0.350/3 = 0.883$. By Eq. 23, $f_c = 2M/(kjbd^2) = 2(744,000)/[0.350(0.883)(12)(19.5)^2] = 1060$ lb/sq.in. (7308.7 kPa). By Eq. 25, $f_s = M/(A_s jd) = 744,000/[2.20(0.883)(19.5)] = 19,600$ lb/sq.in. (135,142 kPa).

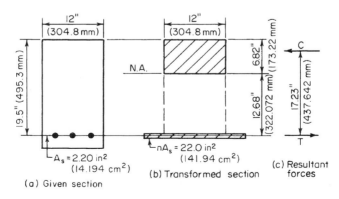

FIGURE 13

CAPACITY OF A RECTANGULAR BEAM

The beam in Fig. 14a is made of 2500-lb/sq.in. (17,237.5-kPa) concrete. Determine the flexural capacity of the member (a) without applying the basic equations of reinforced-concrete beam design; (b) by applying the basic equations.

Calculation Procedure:

1. Record the pertinent beam data
Thus, $f'_c = 2500$ lb/sq.in. (17,237.5 kPa); $\therefore f_{c,\text{allow}} = 1125$ lb/sq.in. (7756.9 kPa); $n = 10$; $A_s = 3.95$ sq.in. (25.485 cm^2); $nA_s = 39.5$ sq.in. (254.85 cm^2).

2. Locate the centroidal axis of the transformed section
Thus, $16(kd)^2/2 - 39.5(23.5 - kd) = 0$; $kd = 8.58$ in. (217.93 mm); $d - kd = 14.92$ in. (378.968 mm).

3. Ascertain which of the two allowable stresses governs the capacity of the member
For this purpose, assume that $f_c = 1125$ lb/sq.in. (7756.9 kPa). By proportion, $f_s = 10(1125)(14.92/8.58) = 19,560$ lb/sq.in. (134,866 kPa) < 20,000 lb/sq.in. (137,900 kPa). Therefore, concrete stress governs.

4. Calculate the allowable bending moment
Thus, $jd = 23.5 - 8.58/3 = 20.64$ in. (524.256 mm); $M = Cjd = \frac{1}{2}(1125)(16)(8.58)(20.64) = 1,594,000$ in.·lb (180,090.1 N·m); or $M = Tjd = 3.95(19,560)(20.64) = 1,594,000$ in.·lb (180,090.1 N·m). This concludes part a of the solution. The next step comprises part b.

5. Compute p and compare with p_b to identify the controlling stress
Thus, from Table 1, $p_b = 0.0101$; then $p = A_s/(bd) = 3.95/[16(23.5)] = 0.0105 > p_b$. Therefore, concrete stress governs.

Applying the basic equations in the proper sequence yields $pn = 0.1050$; by Eq. 30, $k = [0.210 + 0.105^2]^{0.5} - 0.105 = 0.365$; by Eq. 24, $M = (\frac{1}{6})(1125)(0.365)(2.635)(16)(23.5)^2 = 1,593,000$ in.·lb (179,977.1 N·m). This agrees closely with the previously computed value of M.

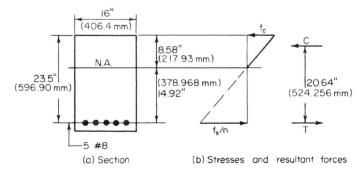

FIGURE 14

DESIGN OF REINFORCEMENT IN A RECTANGULAR BEAM OF GIVEN SIZE

A rectangular beam of 4000-lb/sq.in. (27,580-kPa) concrete has a width of 14 in. (355.6 mm) and an effective depth of 23.5 in. (596.9 mm). Determine the area of reinforcement if the beam is to resist a bending moment of (a) 220 ft·kips (298.3 kN·m); (b) 200 ft·kips (271.2 kN·m).

Calculation Procedure:

1. Calculate the moment capacity of this member at balanced design
Record the following values: $f_{c,\text{allow}} = 1800$ lb/sq.in. (12,411 kPa); $n = 8$. From Table 1, $j_b = 0.860$; $K_b = 324$ lb/sq.in. (2234.0 kPa); $M_b = K_b bd^2 = 324(14)(23.5)^2 = 2,505,000$ in.·lb (283,014.9 N·m).

2. Determine which material will be stressed to capacity under the stipulated moment
For part a, $M = 220,000(12) = 2,640,000$ in.·lb (3,579,840 N·m) $> M_b$. This result signifies that the beam size is deficient with respect to balanced design, and the concrete will therefore be stressed to capacity.

3. Apply the basic equations in proper sequence to obtain A_s
By Eq. 24, $k(3 - k) = 6M/(f_c bd^2) = 6(2,640,000)/[1800(14)(23.5)^2] = 1.138$; $k = 0.446$. By Eq. 29, $p = k^2/[2n(1 - k)] = 0.446^2/[16(0.554)] = 0.0224$; $A_s = pbd = 0.0224(14)(23.5) = 7.37$ sq.in. (47.551 cm^2).

4. Verify the result by evaluating the flexural capacity of the member
For part b, compute A_s by the exact method and then describe the approximate method used in practice.

5. Determine which material will be stressed to capacity under the stipulated moment
Here $M = 200,000(12) = 2,400,000$ in.·lb (3,254,400 N·m) $< M_b$. This result signifies that the beam size is excessive with respect to balanced design, and the steel will therefore be stressed to capacity.

6. Apply the basic equations in proper sequence to obtain A_s
By using Eq. 27, $k^2(3 - k)/(1 - k) = 6nM/(f_s bd^2) = 6(8)(2,400,000)/[20,000(14)(23.5)^2] = 0.7448$; $k = 0.411$. By Eq. 22, $j = 1 - 0.411/3 = 0.863$. By Eq. 25, $A_s = M/(f_s jd) = 2,400,000/[20,000(0.863)(23.5)] = 5.92$ sq.in. (38.196 cm^2).

7. Verify the result by evaluating the flexural capacity of this member
The value of j obtained in step 6 differs negligibly from the value $j_b = 0.860$. Consequently, in those instances where the beam size is only moderately excessive with respect to balanced design, the practice is to consider that $j = j_b$ and to solve Eq. 25 directly on this basis. This practice is conservative, and it obviates the need for solving a cubic equation, thus saving time.

DESIGN OF A RECTANGULAR BEAM

A beam on a simple span of 13 ft (3.9 m) is to carry a uniformly distributed load, exclusive of its own weight, of 3600 lb/lin ft (52,538.0 N/m) and a concentrated load of 17,000 lb (75,616 N) applied at midspan. Design the section, using f'_c = 3000 lb/sq.in. (20,685 kPa).

Calculation Procedure:

1. Record the basic values associated with balanced design
There are two methods of allowing for the beam weight: (a) to determine the bending moment with an estimated beam weight included; (b) to determine the beam size required to resist the external loads alone and then increase the size slightly. The latter method is used here.
From Table 1, K_b = 223 lb/sq.in. (1537.6 kPa); p_b = 0.0128; j_b = 0.874.

2. Calculate the maximum moment caused by the external loads
Thus, the maximum moment M_e = ¼PL + ⅛wL^2 = ¼(17,000)(13)(12) + ⅛(3600)(13)^2(12) = 1,576,000 in.·lb (178,056.4 N·m).

3. Establish a trial beam size
Thus, bd^2 = M/K_b = 1,576,000/223 = 7067 in^3 (115,828.1 cm^3). Setting b = (⅔)d, we find b – 14.7 in. (373.38 mm), d = 22.0 in. (558.8 mm). Try b = 15 in. (381 mm) and d = 22.5 in. (571.5 mm), producing an overall depth of 25 in. (635 mm) if the reinforcing bars may be placed in one row.

4. Calculate the maximum bending moment with the beam weight included; determine whether the trial section is adequate
Thus, beam weight = 15(25)(l50)/144 = 391 lb/lin ft (5706.2 N/m); M_w = (⅛)(391)(13)^2(12) = 99,000 in.·lb (11,185.0 N·m); M = 1,576,000 + 99,000 = 1,675,000 in.·lb (189,241.5 N·m); M_b = $K_b bd^2$ = 223(15)(22.5)^2 = 1,693,000 in.·lb (191,275.1 N·m). The trial section is therefore satisfactory because it has adequate capacity.

5. Design the reinforcement
Since the beam size is slightly excessive with respect to balanced design, the steel will be stressed to capacity under the design load. Equation 25 is therefore suitable for this calculation. Thus, A_s = $M/(f_s jd)$ = 1,675,000/[20,000(0.874)(22.5)] = 4.26 sq.in. (27.485 cm^2).
An alternative method of calculating A_s is to apply the value of p_b while setting the beam width equal to the dimension actually required to produce balanced design. Thus, A_s = 0.0128(15)(1675)(22.5)/1693 = 4.27 sq.in. (27.550 cm^2).
Use one no. 10 and three no. 9 bars, for which A_s = 4.27 sq.in. (27.550 cm^2) and b_{min} = 12.0 in. (304.8 mm).

6. Summarize the design
Thus, beam size is 15 × 25 in. (381 × 635 mm); reinforcement is with one no. 10 and three no. 9 bars.

DESIGN OF WEB REINFORCEMENT

A beam 14 in. (355.6 mm) wide with an 18.5-in. (469.9-mm) effective depth carries a uniform load of 3.8 kips/lin ft (55.46 N/m) and a concentrated midspan load of 2 kips (8.896 kN). The beam is simply supported, and the clear distance between supports is 13 ft

(3.9 m). Using $f'_c = 3000$ lb/sq.in. (20,685 kPa) and an allowable stress f_v in the stirrups of 20,000 lb/sq.in. (137,900 kPa), design web reinforcement in the form of vertical U stirrups.

Calculation Procedure:

1. Construct the shearing-stress diagram for half-span

The design of web reinforcement by the working-stress method parallels the design by the ultimate-strength method, given earlier. Let v = nominal shearing stress, lb/sq.in. (kPa); v'_c = shearing stress resisted by concrete; v' = shearing stress resisted by web reinforcement.

The ACI *Code* provides two alternative methods of computing the shearing stress that may be resisted by the concrete. The simpler method is used here. This sets

$$v_c = 1.1(f'_c)^{0.5} \tag{33}$$

The equation for nominal shearing stress is

$$v = \frac{V}{bd} \tag{34}$$

The shearing-stress diagram for a half-span is shown in Fig. 15. Establish the region AD within which web reinforcement is required. Thus, $v_c = 1.1(3000)^{0.5} = 60$ lb/sq.in. (413.7 kPa). At the face of the support, $V = 6.5(3800) + 1000 = 25,700$ lb (114,313.6 N); $v = 25,700/[14(18.5)] = 99$ lb/sq.in. (682.6 kPa).

At midspan, $V = 1000$ lb (4448 N); $v = 4$ lb/sq.in. (27.6 kPa); slope of diagram = $-(99 - 4)/78 = -1.22$ lb/(in²·in.) (−0.331 kPa/mm). At distance d from the face of the support, $v = 99 - 18.5(1.22) = 76$ lb/sq.in. (524.02 kPa); $v' = 76 - 60 = 16$ lb/sq.in. (110.3 kPa); $AC = (99 - 60)/1.22 = 32$ in. (812.8 mm); $AD = AC + d = 32 + 18.5 = 50.5$ in. (1282.7 mm).

2. Check the beam size for compliance with the Code

Thus, $v_{max} = 5(f'_c)^{0.5} - 274$ lb/sq.in. (1889.23 kPa) > 76 lb/sq.in. (524.02 kPa). This is acceptable.

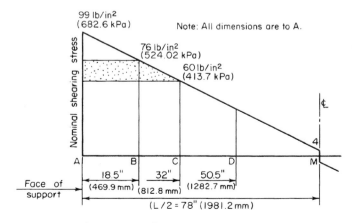

FIGURE 15. Shearing-stress diagram.

3. Select the stirrup size
Use the method given earlier in the ultimate-strength calculation procedure to select the stirrup size, establish the maximum allowable spacing, and devise a satisfactory spacing.

CAPACITY OF A T BEAM

Determine the flexural capacity of the T beam in Fig. 16a, using $f'_c = 3000$ lb/sq.in. (20,685 kPa).

Calculation Procedure:

1. Record the pertinent beam values
The neutral axis of a T beam often falls within the web. However, to simplify the analysis, the resisting moment developed by the concrete lying between the neutral axis and the flange is usually disregarded. Let A_f denote the flange area. The pertinent beam values are $f_{c,\text{allow}} = 1350$ lb/sq.in. (9308.3 kPa); $n = 9$; $k_b = 0.378$; $nA_s = 9(4.00) = 36.0$ sq.in. (232.3 cm^2).

2. Tentatively assume that the neutral axis lies in the web
Locate this axis by taking static moments with respect to the top line. Thus $A_f = 5(16) = 80$ sq.in. (516.2 cm^2); $kd = [80(2.5) + 36.0(21.5)]/(80 + 36.0) = 8.40$ in. (213.36 mm).

3. Identify the controlling stress
Thus $k = 8.40/21.5 = 0.391 > k_b$; therefore, concrete stress governs.

4. Calculate the allowable bending moment
Using Fig. 16c, we see $f_{c1} = 1350(3.40)/8.40 = 546$ lb/sq.in. (3764.7 kPa); $C = \frac{1}{2}(80)(1350 + 546) = 75,800$ lb (337,158.4 N). The action line of this resultant force lies at the centroidal axis of the stress trapezoid. Thus, $z = (5/3)(1350 + 2 \times 546)/(1350 + 546) = 2.15$ in. (54.61 mm); or $z = (5/3)(8.40 + 2 \times 3.40)/(8.40 + 3.40) = 2.15$ in. (54.61 mm); $M = Cjd = 75,800(19.35) = 1,467,000$ in.·lb (165,741 N·m).

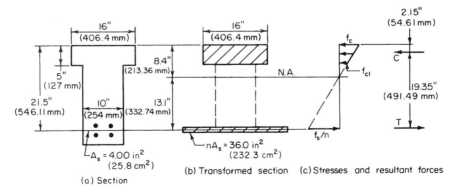

FIGURE 16

5. Alternatively, calculate the allowable bending moment by assuming that the flange extends to the neutral axis

Then apply the necessary correction. Let C_1 = resultant compressive force if the flange extended to the neutral axis, lb (N); C_2 = resultant compressive force in the imaginary extension of the flange, lb (N). Then C_1 = ½(1350)(16)(8.40) = 90,720 lb (403,522.6 N); C_2 = 90,720(3.40/8.40)² = 14,860 lb (66,097.3 N); M = 90,720(21.5 − 8.40/3) − 14,860(21.5 − 5 − 3.40/3) = 1,468,000 in.·lb (165,854.7 N·m).

DESIGN OF A T BEAM HAVING CONCRETE STRESSED TO CAPACITY

A concrete girder of 2500-lb/sq.in. (17,237.5-kPa) concrete has a simple span of 22 ft (6.7 m) and is built integrally with a 5-in. (127-mm) slab. The girders are spaced 8 ft (2.4 m) on centers; the overall depth is restricted to 20 in. (508 mm) by headroom requirements. The member carries a load of 4200 lb/lin ft (61,294.4 N/m), exclusive of the weight of its web. Design the section, using tension reinforcement only.

Calculation Procedure:

1. Establish a tentative width of web
Since the girder is built integrally with the slab that it supports, the girder and slab constitute a structural entity in the form of a T beam. The effective flange width is established by applying the criteria given in the ACI *Code*, and the bending stress in the flange is assumed to be uniform across a line parallel to the neutral axis. Let A_f = area of flange sq.in. (cm²); b = width of flange, in. (mm); b' = width of web, in. (mm); t = thickness of flange, in. (mm); s = center-to-center spacing of girders.

To establish a tentative width of web, try b' = 14 in. (355.6 mm). Then the weight of web = 14(15)(150)/144 = 219, say 220 lb/lin ft (3210.7 N/m); w = 4200 + 220 = 4420 lb/lin ft (64,505.0 N/m).

Since two rows of bars are probably required, d = 20 − 3.5 = 16.5 in. (419.1 mm). The critical shear value is $V = w(0.5L − d) = 4420(11 − 1.4) = 42,430$ lb (188,728.7 N); $v = V/b'd = 42,430/[14(16.5)] = 184$ lb/sq.in. (1268.7 kPa). From the *Code*, $v_{max} = 5(f'_c)^{0.5} = 250$ lb/sq.in. (1723.8 kPa). This is acceptable.

Upon designing the reinforcement, consider whether it is possible to reduce the width of the web.

2. Establish the effective width of the flange according to the Code
Thus, ¼L = ¼(22)(12) = 66 in. (1676.4 mm); $16t + b'$ = 16(5) + 14 = 94 in. (2387.6 mm); s = 8(12) = 96 in. (2438.4 mm); therefore b = 66 in. (1676.4 mm).

3. Compute the moment capacity of the member at balanced design
Compare the result with the moment in the present instance to identify the controlling stress. With Fig. 16 as a guide, $k_b d$ = 0.360(16.5) = 5.94 in. (150.876 mm); A_f = 5(66) = 330 sq.in. (2129.2 cm²); f_{c1} = 1125(0.94)/5.94 = 178 lb/sq.in. (1227.3 kPa); $C_b = T_b$ = ½(330)(1125 + 178) = 215,000 lb (956,320 N); $z_b = (5/3)(5.94 + 2 \times 0.94)/(5.94 + 0.94)$ = 1.89 in. (48.0 mm); jd = 14.61 in. (371.094 mm); M_b = 215,000(14.61) = 3,141,000 in.·lb (354,870.2 N·m); $M = (1/8)(4420)(22)^2(12)$ = 3,209,000 in.·lb (362,552.8 N·m).

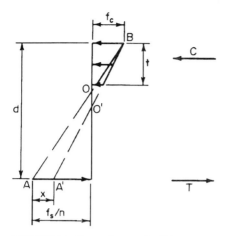

FIGURE 17. Stress diagram for T beam.

The beam size is slightly deficient with respect to balanced design, and the concrete will therefore be stressed to capacity under the stipulated load. In Fig. 17, let AOB represent the stress line associated with balanced design and $A'O'B$ represent the stress line in the present instance. (The magnitude of AA' is exaggerated for clarity.)

4. Develop suitable equations for the beam
Refer to Fig. 17:

$$T = T_b + \frac{bt^2 x}{2d} \qquad (35)$$

where T and T_b = tensile force in present instance and at balanced design, respectively. And

$$M = M_b + \frac{bt^2(3d - 2t)x}{6d} \qquad (36)$$

5. Apply the equations from step 4
Thus, $M - M_b = 68,000$ in.·lb (7682.6 N·m). By Eq. 36, $x = 68,000(6)(16.5)/[66(25)(49.5 - 10)] = 103$ lb/sq.in. (710.2 kPa); $f_s = 20,000 - 10(103) = 18,970$ lb/sq.in. (130,798.2 kPa). By Eq. 35, $T = 215,000 + 66(25)(103)/33 = 220,200$ lb (979,449.6 N).

6. Design the reinforcement; establish the web width
Thus $A_s = 220,200/18,970 = 11.61$ sq.in. (74.908 cm^2). Use five no. 11 and three no. 10 bars, placed in two rows. Then $A_s = 11.61$ sq.in. (74.908 cm^2); $b'_{min} = 14.0$ in. (355.6 mm). It is therefore necessary to maintain the 14-in. (355.6-mm) width.

7. Summarize the design
Width of web: 14 in. (355.6 mm); reinforcement: five no. 11 and three no. 10 bars.

8. Verify the design by computing the capacity of the member
Thus $nA_s = 116.1$ sq.in. (749.08 cm^2); $kd = [330(2.5) + 116.1(16.5)]/(330 + 116.1) = 6.14$ in. (155.956 mm); $k = 6.14/16.5 = 0.372 > k_b$; therefore, concrete is stressed to capacity. Then $f_s = 10(1125)(10.36)/6.14 = 18,980$ lb/sq.in. (130,867.1 kPa); $z = (5/3)(6.14 + 2 \times 1.14)/(6.14 + 1.14) = 1.93$ in. (49.022 mm); $jd = 14.57$ in. (370.078 mm); $M_{allow} = 11.61(18,980)(14.57) = 3,210,000$ in.·lb (362,665.8 N·m). This is acceptable.

REINFORCEMENT FOR DOUBLY REINFORCED RECTANGULAR BEAM

A beam of 4000-lb/sq.in. (27,580-kPa) concrete that will carry a bending moment of 230 ft·kips (311.9 kN·m) is restricted to a 15-in. (381-mm) width and a 24-in. (609.6-mm) total depth. Design the reinforcement.

Calculation Procedure:

1. Record the pertinent beam data
In Fig. 18, where the imposed moment is substantially in excess of that corresponding to balanced design, it is necessary to reinforce the member in compression as well as tension.

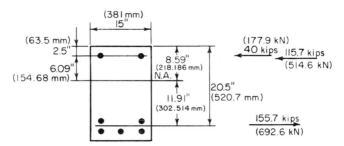

FIGURE 18. Doubly reinforced beam.

The loss in concrete area caused by the presence of the compression reinforcement may be disregarded.

Since plastic flow generates a transfer of compressive stress from the concrete to the steel, the ACI *Code* provides that "in doubly reinforced beams and slabs, an effective modular ratio of 2n shall be used to transform the compression reinforcement and compute its stress, which shall not be taken as greater than the allowable tensile stress." This procedure is tantamount to considering that the true stress in the compression reinforcement is twice the value obtained by assuming a linear stress distribution.

Let A_s = area of tension reinforcement, sq.in. (cm^2); A'_s = area of compression reinforcement, sq.in. (cm^2); f_s = stress in tension reinforcement, lb/sq.in. (kPa); f'_s = stress in compression reinforcement, lb/sq.in. (kPa); C' = resultant force in compression reinforcement, lb (N); M_1 = moment capacity of member if reinforced solely in tension to produce balanced design; M_2 = incremental moment capacity resulting from use of compression reinforcement.

The data recorded for the beam are f_c = 1800 lb/sq.in. (12.411 kPa); n = 8; K_b = 324 lb/sq.in. (2234.0 kPa); k_b = 0.419; j_b = 0.860; M = 230,000(12) = 2,760,000 in.·lb (311,824.8 N·m).

2. Ascertain whether one row of tension bars will suffice
Assume tentatively that the presence of the compression reinforcement does not appreciably alter the value of j. Then jd = 0.860(21.5) = 18.49 in. (469.646 mm); $A_s = M/(f_s jd)$ = 2,760,000/[20,000(18.49)] = 7.46 sq.in. (48.132 cm^2). This area of steel cannot be accommodated in the 15-in. (381-mm) beam width, and two rows of bars are therefore required.

3. Evaluate the moments M_1 and M_2
Thus, d = 24 − 3.5 = 20.5 in. (520.7 mm); $M_1 = K_b bd^2$ = 324(15)(20.5)^2 = 2,040,000 in.·lb (230,479.2 N·m); M_2 = 2,760,000 − 2,040,000 = 720,000 in.·lb (81,345.6 N·m).

4. Compute the forces in the reinforcing steel
For convenience, assume that the neutral axis occupies the same position as it would in the absence of compression reinforcement. For M_1, arm = $j_b d$ = 0.860(20.5) = 17.63 in. (447.802 mm); for M_2, arm = 20.5 − 2.5 = 18.0 in. (457.2 mm); T = 2,040,000/17.63 + 720,000/18.0 = 155,700 lb (692,553.6 N); C' = 40,000 lb (177,920 N).

5. Compute the areas of reinforcement and select the bars
Thus $A_s = T/f_s$ = 155,700/20,000 = 7.79 sq.in. (50.261 cm^2); kd = 0.419(20.5) = 8.59 in. (218.186 mm); $d − kd$ = 11.91 in. (302.514 mm). By proportion, f'_s = 2(20,000)(6.09)/11.91 = 20,500 lb/sq.in. (141,347.5 kPa); therefore, set f'_s = 20,000 lb/sq.in. (137,900 kPa). Then, $A'_s = C'/f'_s$ = 40,000/20,000 = 2.00 sq.in. (12.904 cm^2). Thus tension steel: five no. 11 bars, A_s = 7.80 sq.in. (50.326 cm^2); compression steel: two no. 9 bars, A_s = 2.00 sq.in. (12.904 cm^2).

DEFLECTION OF A CONTINUOUS BEAM

The continuous beam in Fig. 19a and b carries a total load of 3.3 kips/lin ft (48.16 kN/m). When it is considered as a T beam, the member has an effective flange width of 68 in. (1727.2 mm). Determine the deflection of the beam upon application of full live load, using $f_c' = 2500$ lb/sq.in. (17,237.5 kPa) and $f_y = 40,000$ lb/sq.in. (275,800 kPa).

Calculation Procedure:

1. Record the areas of reinforcement
At support: $A_s = 4.43$ sq.in. (28.582 cm^2) (top); $A_s' = 1.58$ sq.in. (10.194 cm^2) (bottom). At center: $A_s = 3.16$ sq.in. (20.388 cm^2) (bottom).

2. Construct the bending-moment diagram
Apply the ACI equation for maximum midspan moment. Refer to Fig. 19c: $M_1 = (1/8)wL'^2 = (1/8)3.3(22)^2 - 200$ ft·kips (271.2 kN·m); $M_2 = (1/16)WL'^2 = 100$ ft·kips (135.6 kN·m); $M_3 = 100$ ft·kips (135.6 kN·m).

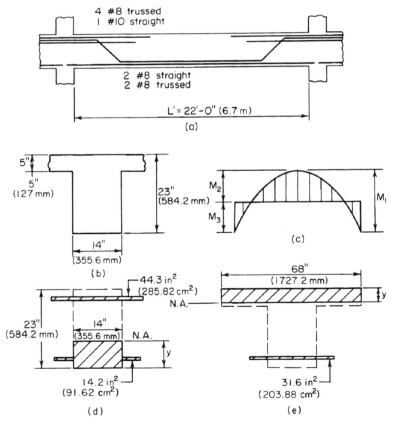

FIGURE 19

3. Determine upon what area the moment of inertia should be based

Apply the criterion set forth in the ACI *Code* to determine whether the moment of inertia is to be based on the transformed gross section or the transformed cracked section. At the support $pf_y = 4.43(40,000)/[14(20.5)] = 617 > 500$. Therefore, use the cracked section.

4. Determine the moment of inertia of the transformed cracked section at the support

Refer to Fig. 19d: $nA_s = 10(4.43) = 44.3$ sq.in. (285.82 cm²); $(n-1)A_s' = 9(1.58) = 14.2$ sq.in. (91.62 cm²). The static moment with respect to the neutral axis is $Q = -\frac{1}{2}(14y^2) + 44.3(20.5 - y) - 14.2(y - 2.5) = 0$; $y = 8.16$ in. (207.264 mm). The moment of inertia with respect to the neutral axis is $I_1 = (\frac{1}{3})14(8.16)^3 + 14.2(8.16 - 2.5)^2 + 44.3(20.5 - 8.16)^2 = 9737$ in⁴ (40.53 dm⁴).

5. Calculate the moment of inertia of the transformed cracked section at the center

Referring to Fig. 19e and assuming tentatively that the neutral axis falls within the flange, we see $nA_s = 10(3.16) = 31.6$ sq.in. (203.88 cm²). The static moment with respect to the neutral axis is $Q = \frac{1}{2}(68y^2) - 31.6(20.5 - y) = 0$; $y = 3.92$ in. (99.568 mm). The neutral axis therefore falls within the flange, as assumed. The moment of inertia with respect to the neutral axis is $I_2 = (\frac{1}{3})68(3.92)3 + 31.6(20.5 - 3.92)^2 = 10,052$ in⁴ (41.840 dm⁴).

6. Calculate the deflection at midspan

Use the equation

$$\Delta = \frac{L'^2}{EI}\left(\frac{5M_1}{48} - \frac{M_3}{8}\right) \tag{37}$$

where I = average moment of inertia, in⁴ (dm⁴). Thus, $I = \frac{1}{2}(9737 + 10,052) = 9895$ in⁴ (41.186 dm⁴); $E = 145^{1.5} \times 33f_c')^{0.5} = 57,600(2500)^{0.5} = 2,880,000$ lb/sq.in. (19,857.6 MPa). Then $\Delta = [22^2 \times 1728/(2880 \times 9895)](5 \times 200/48 - 100/8) = 0.244$ in. (6.198 mm).

Where the deflection under sustained loading is to be evaluated, it is necessary to apply the factors recorded in the ACI *Code*.

Design of Compression Members by Ultimate-Strength Method

The notational system is P_u = ultimate axial compressive load on member, lb (N); P_b = ultimate axial compressive load at balanced design, lb (N); P_0 = allowable ultimate axial compressive load in absence of bending moment, lb (N); M_u = ultimate bending moment in member, lb.·in (N·m); M_b = ultimate bending moment at balanced design; d' = distance from exterior surface to centroidal axis of adjacent row of steel bars, in. (mm); t = overall depth of rectangular section or diameter of circular section, in. (mm).

A compression member is said to be *spirally reinforced* if the longitudinal reinforcement is held in position by spiral hooping and *tied* if this reinforcement is held by means of intermittent lateral ties.

The presence of a bending moment in a compression member reduces the ultimate axial load that the member may carry. In compliance with the ACI *Code*, it is necessary to design for a minimum bending moment equal to that caused by an eccentricity of $0.05t$ for spirally reinforced members and $0.10t$ for tied members. Thus, every compression member that is designed by the ultimate-strength method must be treated as a beam column. This type of member is considered to be in balanced design if failure would be characterized by the simultaneous crushing of the concrete, which is assumed to occur when $\varepsilon_c = 0.003$, and incipient yielding of the tension steel, which occurs when $f_s = f_y$. The ACI Code set $\phi = 0.75$ for spirally reinforced members and $\phi = 0.70$ for tied members.

ANALYSIS OF A RECTANGULAR MEMBER BY INTERACTION DIAGRAM

A short tied member having the cross section shown in Fig. 20a is to resist an axial load and a bending moment that induces compression at A and tension at B. The member is made of 3000-lb/sq.in. (20,685-kPa) concrete, and the steel has a yield point of 40,000 lb/sq.in. (275,800 kPa). By starting with $c = 8$ in. (203.2 mm) and assigning progressively higher values to c, construct the interaction diagram for this member.

Calculation Procedure:

1. Compute the value of c associated with balanced design
An interaction diagram, as the term is used here, is one in which every point on the curve represents a set of simultaneous values of the ultimate moment and allowable ultimate axial load. Let ε_A and ε_B = strain of reinforcement at A and B, respectively; ε_c = strain of extreme fiber of concrete; F_A and F_B = stress in reinforcement at A and B, respectively, lb/sq.in. (kPa); F_A and F_B = resultant force in reinforcement at A and B, respectively; F_c = resultant force in concrete, lb (N).

Compression will be considered positive and tension negative. For simplicity, disregard the slight reduction in concrete area caused by the steel at A.

Referring to Fig. 20b, compute the value of c associated with balanced design. Computing P_b and M_b yields $c_b/d = 0.003/(0.003 + f_y/E_s) = 87,000/(87,000 + f_y)$; $c_b = 10.62$ in. (269.748 mm). Then $\varepsilon_A/\varepsilon_B = (10.62 - 2.5)/(15.5 - 10.62) > 1$; therefore, $f_A = f_y$; $a_b = 0.85(10.62) = 9.03$ in. (229.362 mm); $F_c = 0.85(3000)(12a_b) = 276,300$ lb (1,228,982.4 N); $F_A = 40,000(2.00) = 80,000$ lb (355,840 N); $F_B = -80,000$ lb (−355,840 N); $P_b = 0.70(276,300) = 193,400$ lb (860,243.2 N). Also,

$$M_b = 0.70\left[\frac{F_c(t-a)}{2} + \frac{(F_A - F_B)(t - 2d')}{2}\right] \tag{38}$$

Thus, $M_b = 0.70[276,300(18 - 9.03)/2 + 160,000(6.5)] = 1,596,000$ in.·lb (180,316.1 N·m).

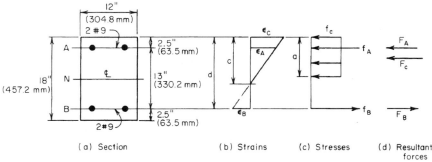

(a) Section   (b) Strains   (c) Stresses   (d) Resultant forces

FIGURE 20

When $c > c_b$, the member fails by crushing of the concrete; when $c < c_b$, it fails by yielding of the reinforcement at line B.

2. Compute the value of c associated with incipient yielding of the compression steel

Compute the corresponding values of P_u and M_u. Since ε_A and ε_B are numerically equal, the neutral axis lies at N. Thus, $c = 9$ in. (228.6 mm); $a = 0.85(9) = 7.65$ in. (194.31 mm); $F_c = 30,600(7.65) = 234,100$ lb (1,041,276.8 N); $F_A = 80,000$ lb (355,840 N); $F_B = -80,000$ lb $(-355,840$ N); $P_u = 0.70$ (234,100) $= 163,900$ lb (729,027.2 N); $M_u = 0.70(234,100 \times 5.18 + 160,000 \times 6.5) = 1,577,000$ in.·lb (178,169.5 N·m).

3. Compute the minimum value of c at which the entire concrete area is stressed to 0.85f'_c

Compute the corresponding values of P_u and M_u. Thus, $a = t = 18$ in. (457.2 mm); $c = 18/0.85 = 21.8$ in. (537.972 mm); $f_B = \varepsilon_c E_s(c-d)/c = 87,000(21.18 - 15.5)/21.18 = 23,300$ lb/sq.in. (160,653.5 kPa); $F_c = 30,600(18) = 550,800$ lb (2,449,958.4 N); $F_A = 80,000$ lb (355,840 N); $F_B = 46,600$ lb (207,276.8 N); $P_u = 0.70(550,800 + 80,000 + 46,600) = 474,200$ lb (2,109,241.6 N); $M_u = 0.70(80,000 - 46,600)6.5 = 152,000$ in.·lb (17,192.9 N·m).

4. Compute the value of c at which $M_u = 0$; compute P_0

The bending moment vanishes when F_B reaches 80,000 lb (355,840 N). From the calculation in step 3, $f_b = 87,000(c-d)/c = 40,000$ lb/sq.in. (275,800 kPa); therefore, $c = 28.7$ in. (728.98 mm); $P_o = 0.70(550,800 + 160,000) = 497,600$ lb (2,213,324.8 N).

5. Assign other values to c, and compute P_u and M_u

By assigning values to c ranging from 8 to 28.7 in. (203.2 to 728.98 mm), typical calculations are: when $c = 8$ in. (203.2 mm); $f_B = -40,000$ lb/sq.in. (−275,800 kPa); $f_A = 40,000(5.5/7.5) = 29,300$ lb/sq.in. (202,023.5 kPa); $a = 6.8$ in. (172.72 mm); $F_c = 30,600(6.8) = 208,100$ lb (925,628.8 N); $P_u = 0.70(208,100 + 58,600 - 80,000) = 130,700$ lb (581,353.6 N); $M_u = 0.70$ (208,100 × 5.6 + 138,600 × 6.5) = 1,446,000 in.·lb (163,369.1 N·m).

When $c = 10$ in. (254 mm), $f_A = 40,000$ lb/sq.in. (275,800 kPa); $f_B = -40,000$ lb/sq.in. (−275,800 kPa); $a = 8.5$ in. (215.9 mm); $F_c = 30,600(8.5) = 260,100$ lb (1,156,924.8 N); $P_u = 0.70(260,100) = 182,100$ lb (809,980 N); $M_u = 0.70(260,100 \times 4.75 + 160,000 \times 6.5) = 1,593,000$ in.·lb (179,997.1 N·m).

When $c = 14$ in. (355.6 mm), $f_B = 87,000(14 - 15.5)/14 = -9320$ lb/sq.in. (−64,261.4 kPa); $a = 11.9$ in. (302.26 mm); $F_c = 30,600(11.9) = 364,100$ lb (1,619,516.8 N); $P_u = 0.70(364,100 + 80,000 - 18,600) = 297,900$ lb (1,325,059.2 N); $M_u = 0.70(364,100 \times 3.05 + 98,600 \times 6.5) = 1,226,000$ in.·lb (138,513.5 N·m).

6. Plot the points representing computed values of P_u and M_u in the interaction diagram

Figure 21 shows these points. Pass a smooth curve through these points. Note that when $P_u < P_b$, a reduction in M_u is accompanied by a reduction in the allowable load P_u.

AXIAL-LOAD CAPACITY OF RECTANGULAR MEMBER

The member analyzed in the previous calculation procedure is to carry an eccentric longitudinal load. Determine the allowable ultimate load if the eccentricity as measured from N is (a) 9.2 in. (233.68 mm); (b) 6 in. (152.4 mm).

Calculation Procedure:

1. Evaluate the eccentricity associated with balanced design

Let e denote the eccentricity of the load and e_b the eccentricity associated with balanced design. Then $M_u = P_u e$. In Fig. 21, draw an arbitrary radius vector OD; then $\tan \theta = ED/OE =$ eccentricity corresponding to point D.

Proceeding along the interaction diagram from A to C, we see that the value of c increases and the value of e decreases. Thus, c and e vary in the reverse manner. To evaluate the allowable loads, it is necessary to identify the portion of the interaction diagram to which each eccentricity applies.

From the computations of the previous calculation procedure, $e_b = M_b/P_b = 1,596,000/193,400 = 8.25$ in. (209.55 mm). This result discloses that an eccentricity of 9.2 in. (233.68 mm) corresponds to a point on AB and an eccentricity of 6 in. (152.4 mm) corresponds to a point on BC.

2. Evaluate P_u when $e = 9.2$ in. (233.68 mm)

It was found that $c = 9$ in. (228.6 mm) is a significant value. The corresponding value of e is $1,577,000/163,900 = 9.62$ in. (244.348 mm). This result discloses that in the present instance $c > 9$ in. (228.6 mm) and consequently $f_A = f_y$; $F_A = 80,000$ lb (355,840 N); $F_B = -80,000$ lb (−355,840 N); $F_c = 30,600a$; $P_u/0.70 = 30,600a$; $M_u/0.70 = 30,600a(18 - a)/2 + 160,000(6.5)$; $e = M_u/P_u = 9.2$ in. (233.68 mm). Solving gives $a = 8.05$ in. (204.47 mm), $P_u = 172,400$ lb (766,835.2 N).

3. Evaluate P_u when $e = 6$ in. (152.4 mm)

To simplify this calculation, the ACI *Code* permits replacement of curve BC in the interaction diagram with a straight line through B and C. The equation of this line is

$$P_u = P_o - (P_o - B_b)\frac{M_u}{M_b} \tag{39}$$

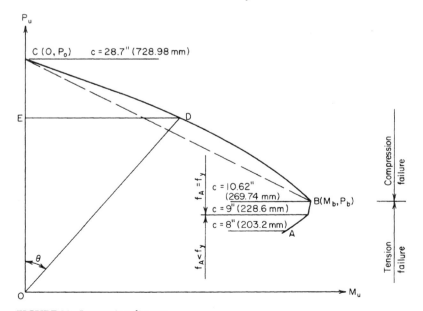

FIGURE 21. Interaction diagram.

By replacing M_u with $P_u e$, the following relation is obtained:

$$P_u = \frac{P_o}{1+(P_o - P_b)e/M_b} \tag{39a}$$

In the present instance, $P_o = 497{,}600$ lb (2,213,324.8 N); $p_b = 193{,}400$ lb (860,243.2 N); $M_b = 1{,}596{,}000$ in.·lb (180,316.1 N·m). Thus $P_u = 232{,}100$ lb (1,032,380 N).

Design of Compression Members by Working-Stress Method

The notational system is as follows: A_g = gross area of section, sq.in. (cm^2); A_s = area of tension reinforcement, sq.in. (cm^2); A_{st} = total area of longitudinal reinforcement, sq.in. (cm^2); D = diameter of circular section, in. (mm); $p_g = A_{st}/A_g$; P = axial load on member, lb (N); f_s = allowable stress in longitudinal reinforcement, lb/sq.in. (kPa); $m = f_y/(0.85 f_c')$.

The working-stress method of designing a compression member is essentially an adaptation of the ultimate-strength method. The allowable ultimate loads and bending moments are reduced by applying an appropriate factor of safety, and certain simplifications in computing the ultimate values are introduced.

The allowable concentric load on a short spirally reinforced column is $P = A_g(0.25 f_c' + f_s p_g)$, or

$$P = 0.25 f_c' A_g + f_s A_{st} \tag{40}$$

where $f_s = 0.40 f_y$, but not to exceed 30,000 lb/sq.in. (206,850 kPa).

The allowable concentric load on a short tied column is $P = 0.85 A_g (0.25 f_c' + f_s p_g)$, or

$$P = 0.2125 f_c' A_g + 0.85 f_s A_{st} \tag{41}$$

A section of the ACI *Code* provides that P_g may range from 0.01 to 0.08. However, in the case of a circular column in which the bars are to be placed in a single circular row, the upper limit of P_g is often governed by clearance. This section of the *Code* also stipulates that the minimum bar size to be used is no. 5 and requires a minimum of six bars for a spirally reinforced column and four bars for a tied column.

DESIGN OF A SPIRALLY REINFORCED COLUMN

A short circular column, spirally reinforced, is to support a concentric load of 420 kips (1868.16 kN). Design the member, using $f_c' = 4000$ lb/sq.in. (27,580 kPa) and $f_y = 50{,}000$ lb/sq.in. (344,750 kPa).

Calculation Procedure:

1. Assume $p_g = 0.025$ and compute the diameter of the section

Thus, $0.25 f_c' = 1000$ lb/sq.in. (6895 kPa); $f_s = 20{,}000$ lb/sq.in. (137,900 kPa). By Eq. 40, $A_g = 420/(1 + 20 \times 0.025) = 280$ sq.in. (1806.6 cm^2). Then $D = (A_g/0.785)^{0.5} = 18.9$ in. (130.32 mm). Set $D = 19$ in. (131.01 mm), making $A_g = 283$ sq.in. (1825.9 cm^2).

2. Select the reinforcing bars

The load carried by the concrete = 283 kips (1258.8 kN). The load carried by the steel = 420 − 283 = 137 kips (609.4 kN). Then the area of the steel is A_{st}, = 137/20 = 6.85 sq.in. (44.196 cm²). Use seven no. 9 bars, each having an area of 1 sq.in. (6.452 cm²). Then A_{st}, = 7.00 sq.in. (45.164 cm²). The *Reinforced Concrete Handbook* shows that a 19-in. (482.6-mm) column can accommodate 11 no. 9 bars in a single row.

3. Design the spiral reinforcement

The portion of the column section bounded by the outer circumference of the spiral is termed the *core* of the section. Let A_c = core area, sq.in. (cm²); D_c = core diameter, in. (mm); a = cross-sectional area of spiral wire, sq.in. (cm²); g = pitch of spiral, in. (mm); p_s = ratio of volume of spiral reinforcement to volume of core.

The ACI *Code* requires 1.5-in. (38.1-mm) insulation for the spiral, with g restricted to a maximum of $D_c/6$. Then D_c = 19 − 3 = 16 in. (406.4 mm); A_c = 201 sq.in. (1296.9 cm²); $D_c/6$ = 2.67 in. (67.818 mm). Use a 2.5-in. (63.5-mm) spiral pitch. Taking a 1-in. (25.4-mm) length of column,

$$p_s = \frac{\text{volume of spiral}}{\text{volume of core}} = \frac{a_s \pi D_c / g}{\pi D_c^2 / 4}$$

or

$$a_s = \frac{g D_c p_s}{4} \quad (42)$$

The required value of p_s, as given by the ACI *Code* is

$$p_s = \frac{0.45(A_g/A_c - 1)f'_c}{f_y} \quad (43)$$

or p_s = 0.45(283/201 − 1)4/50 = 0.0147; a_s = 2.5(16)(0.0147)/4 = 0.147 sq.in. (0.9484 cm²). Use ½-in. (12.7-mm) diameter wire with a_s = 0.196 sq.in. (1.2646 cm²).

4. Summarize the design

Thus: column size: 19-in. (482.6-mm) diameter; longitudinal reinforcement: seven no. 9 bars; spiral reinforcement: ½-in. (12.7-mm) diameter wire, 2.5-in. (63.5-mm) pitch.

ANALYSIS OF A RECTANGULAR MEMBER BY INTERACTION DIAGRAM

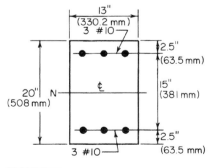

FIGURE 22

A short tied member having the cross section shown in Fig. 22 is to resist an axial load and a bending moment that induces rotation about axis N. The member is made of 4000-lb/sq.in. (27,580-kPa) concrete, and the steel has a yield point of 50,000 lb/sq.in. (344,750 kPa). Construct the interaction diagram for this member.

Calculation Procedure:

1. Compute a and M

Consider a composite member of two materials having equal strength in tension and compression, the member being subjected to an axial load P and bending moment M that induce the allowable stress in one or both materials. Let P_a = allowable axial load in absence of bending moment, as computed by dividing the allowable ultimate load by a factor of safety; M_f = allowable bending moment in absence of axial load, as computed by dividing the allowable ultimate moment by a factor of safety.

Find the simultaneous allowable values of P and M by applying the interaction equation

$$\frac{P}{P_u}+\frac{M}{M_f}=1 \tag{44}$$

Alternative forms of this equation are

$$M=M_f\left(1-\frac{P}{P_a}\right) \qquad P=P_a\left(1-\frac{M}{M_f}\right) \tag{44a}$$

$$P=\frac{P_a M_f}{M_f+P_a M/P} \tag{44b}$$

Equation 44 is represented by line AB in Fig. 23; it is also valid with respect to a reinforced-concrete member for a certain range of values of P and M. This equation is not applicable in the following instances: (*a*) If M is relatively small, Eq. 44 yields a value of P in excess of that given by Eq. 41. Therefore, the interaction diagram must contain line CD, which represents the maximum value of P.

(*b*) If M is relatively large, the section will crack, and the equal-strength assumption underlying Eq. 44 becomes untenable.

Let point E represent the set of values of P and M that will cause cracking in the extreme concrete fiber. And let P_b = axial load represented by point E; M_b = bending moment represented by point E; M_o = allowable bending moment in reinforced-concrete member in absence of axial load, as computed by dividing the allowable ultimate moment by a factor

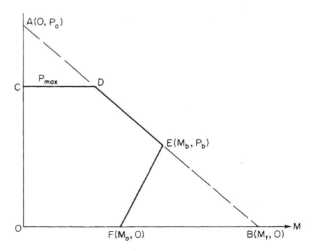

FIGURE 23. Interaction diagram.

of safety. (M_o differs from M_f in that the former is based on a cracked section and the latter on an uncracked section. The subscript b as used by the ACI *Code* in the present instance does *not* refer to balanced design. However, its use illustrates the analogy with ultimate-strength analysis.) Let F denote the point representing M_o.

For simplicity, the interaction diagram is assumed to be linear between E and F. The interaction equation for a cracked section may therefore be expressed in any of the following forms:

$$M = M_o + \left(\frac{P}{P_b}\right)(M_b - M_o) \qquad P = P_b\left(\frac{M - M_o}{M_b - M_o}\right) \tag{45a}$$

$$P = \frac{P_b M_o}{M_o - M_b + P_b M/P} \tag{45b}$$

The ACI *Code* gives the following approximations: For spiral columns:

$$M_o = 0.12 A_{st} f_y D_s \tag{46a}$$

where D_s = diameter of circle through center of longitudinal reinforcement. For symmetric tied columns:

$$M_o = 0.40 A_s f_y (d - d') \tag{46b}$$

For unsymmetric tied columns:

$$M_o = 0.40 A_s f_y jd \tag{46c}$$

For symmetric spiral columns:

$$\frac{M_b}{P_b} = 0.43 p_g m D_s + 0.14 t \tag{47a}$$

For symmetric tied columns:

$$\frac{M_b}{P_b} = d(0.67 p_g m + 0.17) \tag{47b}$$

For unsymmetric tied columns:

$$\frac{M_b}{P_b} = \frac{p'm(d-d') + 0.1d}{(p'-p)m + 0.6} \tag{47c}$$

where p' = ratio of area of compression reinforcement to effective area of concrete. The value of P_a is taken as

$$P_a = 0.34 f'_c A_g (1 + p_g m) \tag{48}$$

The value of M_f is found by applying the section modulus of the transformed uncracked section, using a modular ratio of $2n$ to account for stress transfer between steel and concrete engendered by plastic flow. (If the steel area is multiplied by $2n - 1$, allowance is made for the reduction of the concrete area.)

Computing P_a and M_f yields $A_g = 260$ sq.in. (1677.5 cm²); $A_{st} = 7.62$ sq.in. (49.164 cm²); $p_g = 7.62/260 = 0.0293$; $m = 50/[0.85(4)] = 14.7$; $p_g m = 0.431$; $n = 8$; $P_a = 0.34(4)(260)$ (1.431) = 506 kips (2250.7 kN).

The section modulus to be applied in evaluating M_f is found thus: $I = (1/12)(13)(20)^3 + 7.62(15)(7.5)^2 = 15,100$ in⁴ (62.85 dm⁴); $S = I/c = 15,100/10 = 1510$ in³ (24,748.9 cm³); $M_f = Sf_c = 1510(1.8) = 2720$ in.·kips (307.3 kN·m).

2. Compute P_b and M_b
By Eq. 47b, $M_b/P_b = 17.5(0.67 \times 0.431 + 0.17) = 8.03$ in. (203.962 mm). By Eq. 44b, $P_b = P_a M_f/(M_j + 8.03 P_a) = 506 \times 2720/(2720 + 8.03 \times 506) = 203$ kips (902.9 kN); $M_b = 8.03(203) = 1630$ in.·kips (184.2 kN·m).

3. Compute M_o
By Eq. 46b, $M_o = 0.40(3.81)(50)(15) = 1140$ in.·kips (128.8 kN·m).

4. Compute the limiting value of P
As established by Eq. 41, $P_{max} = 0.2125(4)(260) + 0.85(20)(7.62) = 351$ kips (1561.2 kN).

5. Construct the interaction diagram
The complete diagram is shown in Fig. 23.

AXIAL-LOAD CAPACITY OF A RECTANGULAR MEMBER

The member analyzed in the previous calculation procedure is to carry an eccentric longitudinal load. Determine the allowable load if the eccentricity as measured from N is (*a*) 10 in. (254 mm); (*b*) 6 in. (152.4 mm).

Calculation Procedure:

1. Evaluate P when e = 10 in. (254 mm)
As the preceding calculations show, the eccentricity corresponding to point E in the interaction diagram is 8.03 in. (203.962 mm). Consequently, an eccentricity of 10 in. (254 mm) corresponds to a point on EF, and an eccentricity of 6 in. (152.4 mm) corresponds to a point on ED.

By Eq. 45b, $P = 203(1140)7(1140 - 1630 + 203 \times 10) = 150$ kips (667.2 kN).

2. Evaluate P when e = 6 in. (152.4 mm)
By Eq. 44b, $P = 506(2720)/(2720 + 506 \times 6) = 239$ kips (1063.1 kN).

Design of Column Footings

A reinforced-concrete footing supporting a single column differs from the usual type of flexural member in the following respects: It is subjected to bending in all directions, the ratio of maximum vertical shear to maximum bending moment is very high, and it carries a

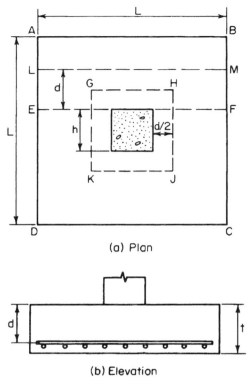

FIGURE 24

heavy load concentrated within a small area. The consequences are as follows: The footing requires two-way reinforcement, its depth is determined by shearing rather than bending stress, the punching-shear stress below the column is usually more critical than the shearing stress that results from ordinary beam action, and the design of the reinforcement is controlled by the bond stress as well as the bending stress.

Since the footing weight and soil pressure are collinear, the former does not contribute to the vertical shear or bending moment. It is convenient to visualize the footing as being subjected to an upward load transmitted by the underlying soil and a downward reaction supplied by the column, this being, of course, an inversion of the true form of loading. The footing thus functions as an overhanging beam. The effective depth of footing is taken as the distance from the top surface to the center of the upper row of bars, the two rows being made identical to avoid confusion.

Refer to Fig. 24, which shows a square footing supporting a square, symmetrically located concrete column. Let P = column load, kips (kN); p = net soil pressure (that caused by the column load alone), lb/sq.ft. (kPa); A = area of footing, sq/ft. (m²); L = side of footing, ft (m); h = side of column, in (mm); d = effective depth of footing, ft (m); t = thickness of footing, ft (m); f_b = bearing stress at interface of column, lb/sq.in. (kPa); v_1 = nominal shearing stress under column, lb/sq.in. (kPa); v_2 = nominal shearing stress caused by beam action, lb/sq.in. (kPa); b_o = width of critical section for v_1, ft (m); V_1 and V_2 = vertical shear at critical section for stresses v_1 and v_2, respectively.

In accordance with the ACI *Code*, the critical section for v_1 is the surface *GHJK*, the sides of which lie at a distance $d/2$ from the column faces. The critical section for v_2 is plane *LM*, located at a distance d from the face of the column. The critical section for bending stress and bond stress is plane *EF* through the face of the column. In calculating v_2, f, and u, no allowance is made for the effects of the orthogonal reinforcement.

DESIGN OF AN ISOLATED SQUARE FOOTING

A 20-in (508-mm) square tied column reinforced with eight no. 9 bars carries a concentric load of 380 kips (1690.2 kN). Design a square footing by the working-stress method using these values: the allowable soil pressure is 7000 lb/sq.ft. (335.2 kPa); f'_c = 3000 lb/sq.in. (20,685 kPa); and f_s = 20,000 lb/sq.in. (137,900 kPa).

Calculation Procedure:

1. Record the allowable shear, bond, and bearing stresses
From the ACI *Code* table, $v_1 = 110$ lb/sq.in. (758.5 kPa); $v_2 = 60$ lb/sq.in. (413.7 kPa); $f_b = 1125$ lb/sq.in. (7756.9 kPa); $u = 4.8(f'_c)^{0.5}$/bar diameter = 264/bar diameter.

2. Check the bearing pressure on the footing
Thus, $f_{b'} = 380/[20(20)] = 0.95$ kips/sq.in. (7.258 MPa) < 1.125 kips/sq.in. (7.7568 MPa). This is acceptable.

3. Establish the length of footing
For this purpose, assume the footing weight is 6 percent of the column load. Then $A = 1.06(380)/7 = 57.5$ sq.ft. (5.34 sq.in.). Make $L = 7$ ft 8 in. $= 7.67$ ft (2.338 m); $A = 58.8$ sq.ft. (5.46 m²).

4. Determine the effective depth as controlled by v_1
Apply

$$(4v_1 + p)d^2 + h(4v_1 + 2p)d = p(A - h^2) \qquad (49)$$

Verify the result after applying this equation. Thus $p = 380/58.8 = 6.46$ kips/sq.ft. (0.309 MPa) $= 0.11(144) = 15.84$ kips/sq.ft. (0.758 MPa); $69.8d^2 + 127.1d = 361.8$; $d = 1.54$ ft (0.469 m). Checking in Fig. 24, we see $GH = 1.67 + 1.54 = 3.21$ ft (0.978 m); $V_1 = 6.46(58.8 - 3.21^2) = 313$ kips (1392.2 kN); $v_1 = V_1/(b_o d) = 313/[4(3.21)(1.54)] = 15.83$ kips/sq.ft. (0.758 MPa). This is acceptable.

5. Establish the thickness and true depth of footing
Compare the weight of the footing with the assumed weight. Allowing 3 in. (76.2 mm) for insulation and assuming the use of no. 8 bars, we see that $t = d + 4.5$ in. (114.3 mm). Then $t = 1.54(12) + 4.5 = 23.0$ in. (584.2 mm). Make $t = 24$ in. (609.6 mm); $d = 19.5$ in. $= 1.63$ ft (0.496 m). The footing weight = $58.8(2)(0.150) = 17.64$ kips (1384.082 kN). The assumed weight = $0.06(380) = 22.8$ kips (101.41 kN). This is acceptable.

6. Check v_2
In Fig. 24, $AL = (7.67 - 1.67)/2 - 1.63 = 1.37$ ft (0.417 m); $V_2 = 380(1.37/7.67) = 67.9$ kips (302.02 kN); $v_2 = V_2/(Ld) = 67,900/[92(19.5)] = 38$ lb/sq.in. (262.0 kPa) < 60 lb/sq.in. (413.7 kPa). This is acceptable.

7. Design the reinforcement
In Fig. 24, $EA = 3.00$ ft (0.914 m); $V_{EF} = 380(3.00/7.67) = 148.6$ kips (666.97 kN); $M_{EF} = 148.6(½)(3.00)(12) = 2675$ in.·kips (302.22 kN·m); $A_s = 2675/[20(0.874)(19.5)] = 7.85$ sq.in. (50.648 cm²). Try 10 no. 8 bars each way. Then $A_s = 7.90$ sq.in. (50.971 cm²); $\Sigma o = 31.4$ in. (797.56 mm); $u = V_{EF}/\Sigma ojd = 148,600/[31.4(0.874)(19.5)] = 278$ lb/sq.in. (1916.81 kPa); $u_{allow} = 264/1 = 264$ lb/sq.in. (1820.3 kPa).

The bond stress at EF is slightly excessive. However, the ACI *Code*, in sections based on ultimate-strength considerations, permits disregarding the local bond stress if the average bond stress across the length of embedment is less than 80 percent of the allowable stress. Let L_e denote this length. Then $L_e = EA - 3 = 33$ in. (838.2 mm); $0.80 u_{allow} = 211$ lb/sq.in. (1454.8 kPa); $u_{av} = A_s f_s/(L_e \Sigma o) = 0.79(20,000)/[33(3.1)] = 154$ lb/sq.in. (1061.8 kPa). This is acceptable.

8. Design the dowels to comply with the Code
The function of the dowels is to transfer the compressive force in the column reinforcing bars to the footing. Since this is a tied column, assume the stress in the bars is $0.85(20,000) = 17,000$ lb/sq.in. (117,215.0 kPa). Try eight no. 9 dowels with $f_y = 40,000$ lb/sq.in. (275,800.0 kPa). Then $u = 264/(9/8) = 235$ lb/sq.in. (1620.3 kPa); $L_e = 1.00(17,000)/[235(3.5)] = 20.7$ in.

(525.78 mm). Since the footing can provide a 21-in. (533.4-mm) embedment length, the dowel selection is satisfactory. Also, the length of lap = 20(9/8) = 22.5 in. (571.5 mm); length of dowels = 20.7 + 22.5 = 43.2, say 44 in. (1117.6 mm). The footing is shown in Fig. 25.

COMBINED FOOTING DESIGN

An 18-in. (457.2-mm) square exterior column and a 20-in. (508.0-mm) square interior column carry loads of 250 kips (1112 kN) and 370 kips (1645.8 kN), respectively. The column centers are 16 ft (4.9 m) apart, and the footing cannot project beyond the face of the exterior column. Design a combined rectangular footing by the working-stress method, using f'_c = 3000 lb/sq.in. (20,685.0 kPa), f_s = 20,000 lb/sq.in. (137,900.0 kPa), and an allowable soil pressure of 5000 lb/sq.in. (239.4 kPa).

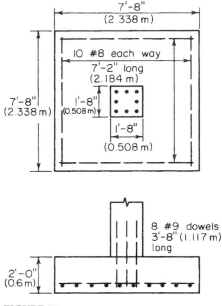

FIGURE 25

Calculation Procedure:

1. Establish the length of footing, applying the criterion of uniform soil pressure under total live and dead loads

In many instances, the exterior column of a building cannot be individually supported because the required footing would project beyond the property limits. It then becomes necessary to use a combined footing that supports the exterior column and the adjacent interior column, the footing being so proportioned that the soil pressure is approximately uniform.

The footing dimensions are shown in Fig. 26a, and the reinforcement is seen in Fig. 27. It is convenient to visualize the combined footing as being subjected to an upward load transmitted by the underlying soil and reactions supplied by the columns. The member thus functions as a beam that overhangs one support. However, since the footing is considerably wider than the columns, there is a transverse bending as well as longitudinal bending in the vicinity of the columns. For simplicity, assume that the transverse bending is confined to the regions bounded by planes AB and EF and by planes GH and NP, the distance m being h/2 or d/2, whichever is smaller.

In Fig. 26a, let Z denote the location of the resultant of the column loads. Then x = 370(16)/(250 + 370) = 9.55 ft (2.910 m). Since Z is to be the centroid of the footing, L = 2(0.75 + 9.55) = 20.60 ft (6.278 m). Set L = 20 ft 8 in. (6.299 m), but use the value 20.60 ft (6.278 m) in the stress calculations.

2. Construct the shear and bending-moment diagrams

The net soil pressure per foot of length = 620/20.60 = 30.1 kips/lin ft (439.28 kN/m). Construct the diagrams as shown in Fig. 26.

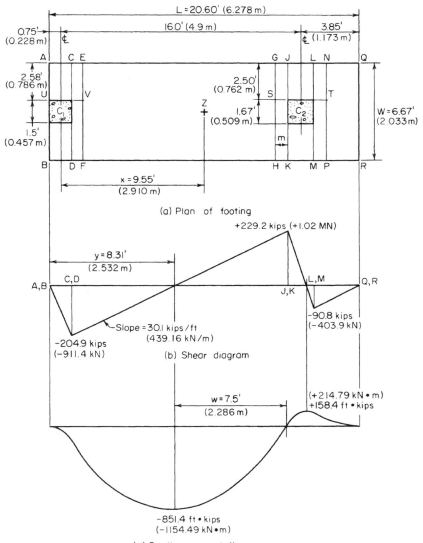

FIGURE 26

3. Establish the footing thickness
Use

$$(Pv_2 + 0.17VL + Pp')d - 0.17Pd^2 - VLp' \quad (50)$$

where P = aggregate column load, kips (kN); V = maximum vertical shear at a column face, kips (kN); p' = gross soil pressure, kips/sq.ft. (MPa).

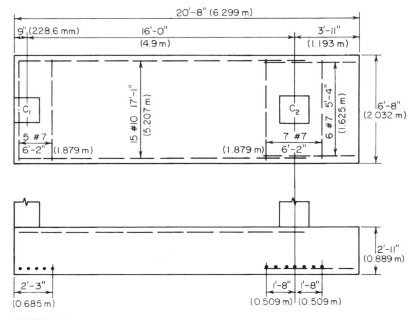

FIGURE 27

Assume that the longitudinal steel is centered 3½ in. (88.9 mm) from the face of the footing. Then $P = 620$ kips (2757.8 kN); $V = 229.2$ kips (1019.48 kN); $v_2 = 0.06(144) = 8.64$ kips/sq.ft. (0.414 MPa); $9260d - 105.4d^2 = 23{,}608$; $d = 2.63$ ft (0.801 m); $t = 2.63 + 0.29 = 2.92$ ft. Set $t = 2$ ft 11 in. (0.889 m); $d = 2$ ft 7½ in. (0.800 m).

4. Compute the vertical shear at distance d from the column face
Establish the width of the footing. Thus $V = 229.2 - 2.63(30.1) = 150.0$ kips (667.2 kN); $v = V/(Wd)$, or $W = V/(vd) = 150/[8.64(2.63)] = 6.60$ ft (2.012 m). Set $W = 6$ ft 8 in.(2.032 m).

5. Check the soil pressure
The footing weight $= 20.67(6.67)(2.92)(0.150) = 60.4$ kips (268.66 kN); $p' = (620 + 60.4)/[(20.67)(6.67)] = 4.94$ kips/sq.ft. (0.236 MPa) $<$ 5 kips/sq.ft. (0.239 MPa). This is acceptable.

6. Check the punching shear
Thus, $p = 4.94 - 2.92(0.150) = 4.50$ kips/sq.ft. (0.215 MPa). At C1: $b_o = 18 + 31.5 + 2(18 + 15.8) = 117$ in. (2971.8 mm); $V = 250 - 4.50(49.5)(33.8)/144 = 198$ kips (880.7 kN); $v_1 = 198{,}000/[117(31.5)] = 54$ lb/sq.in. (372.3 kPa) $<$ 110 lb/sq.in. (758.5 kPa); this is acceptable.

At C2: $b_o = 4(20 + 31.5) = 206$ in. (5232.4 mm); $V = 370 - 4.50(51.5)^2/144 = 287$ kips (1276.6 kN); $v_1 = 287{,}000/[206(31.5)] = 44$ lb/sq.in. (303.4 kPa). This is acceptable.

7. Design the longitudinal reinforcement for negative moment
Thus, $M = 851{,}400$ ft·lb $= 10{,}217{,}000$ in.·lb (1,154,316.6 N·m); $M_b = 223(80)(31.5)^2 = 17{,}700{,}000$ in.·lb (1,999,746.0 N·m). Therefore, the steel is stressed to capacity, and $A_s = 10{,}217{,}000/[20{,}000(0.874)(31.5)] = 18.6$ sq.in. (120.01 cm²). Try 15 no. 10 bars with $A_s = 19.1$ sq.in. (123.2 cm²); $\Sigma o = 59.9$ in. (1521.46 mm).

The bond stress is maximum at the point of contraflexure, where $V = 15.81(30.1) - 250 = 225.9$ kips (1004.80 kN); $u = 225{,}900/[59.9(0.874)(31.5)] = 137$ lb/sq.in. (944.6 kPa); $u_{\text{allow}} = 3.4(3000)^{0.5}/1.25 = 149$ lb/sq.in. (1027.4 kPa). This is acceptable.

8. Design the longitudinal reinforcement for positive moment

For simplicity, design for the maximum moment rather than the moment at the face of the column. Then $A_s = 158{,}400(12)/[20{,}000(0.874)(31.5)] = 3.45$ sq.in. (22.259 cm^2). Try six no. 7 bars with $A_s = 3.60$ sq.in. (23.227 cm^2); $\Sigma o = 16.5$ in. (419.10 mm). Take LM as the critical section for bond, and $u = 90{,}800/[16.5(0.874)(31.5)] = 200$ lb/sq.in. (1379.0 kPa); $u_{allow} = 4.8(3000)^{0.5}/0.875 = 302$ lb/sq.in. (2082.3 kPa). This is acceptable.

9. Design the transverse reinforcement under the interior column

For this purpose, consider member $GNPH$ as an independent isolated footing. Then $V_{ST} = 370(2.50/6.67) = 138.8$ kips (617.38 kN); $M_{ST} = \frac{1}{2}(138.8)(2.50)(12) = 2082$ in.·kips (235.22 kN·m). Assume $d = 35 - 4.5 = 30.5$ in. (774.7 mm); $A_s = 2{,}082{,}000/[20{,}000(0.874)(30.5)] = 3.91$ sq.in. (25.227 cm^2). Try seven no. 7 bars; $A_s = 4.20$ sq.in. (270.098 cm^2); $\Sigma o = 19.2$ in. (487.68 mm); $u = 138{,}800/[19.2(0.874)(30.5)] = 271$ lb/sq.in. (1868.5 kPa); $u_{allow} = 302$ lb/sq.in. (2082.3 kPa). This is acceptable.

Since the critical section for shear falls outside the footing, shearing stress is not a criterion in this design.

10. Design the transverse reinforcement under the exterior column; disregard eccentricity

Thus, $V_{UV} = 250(2.58/6.67) = 96.8$ kips (430.57 kN); $M_{UV} = \frac{1}{2}(96.8)(2.58)(12) = 1498$ in.·kips (169.3 kN·m); $A_s = 2.72$ sq.in. (17.549 cm^2). Try five no. 7 bars; $A_s = 3.00$ sq.in. (19.356 cm^2); $\Sigma o = 13.7$ in. (347.98 mm); $u = 96{,}800/[13.7(0.874)(31.5)] = 257$ lb/sq.in. (1772.0 kPa). This is acceptable.

Cantilever Retaining Walls

Retaining walls having a height ranging from 10 to 20 ft (3.0 to 6.1 m) are generally built as reinforced-concrete cantilever members. As shown in Fig. 28, a cantilever wall comprises a vertical stem to retain the soil, a horizontal base to support the stem, and in many instances a key that projects into the underlying soil to augment the resistance to sliding. Adequate drainage is an essential requirement, because the accumulation of water or ice behind the wall would greatly increase the horizontal thrust.

The calculation of earth thrust in this section is based on Rankine's theory, which is developed in a later calculation procedure. When a live load, termed a *surcharge*, is applied to the retained soil, it is convenient to replace this load with a hypothetical equivalent prism of earth. Referring to Fig. 28, consider a portion QR of the wall, R being at distance y below the top. Take the length of wall normal to the plane of the drawing as 1 ft (0.3 m). Let T = resultant earth thrust on QR; M = moment of this thrust with respect to R; h = height of equivalent earth prism that replaces surcharge; w = unit weight of earth; C_a = coefficient of

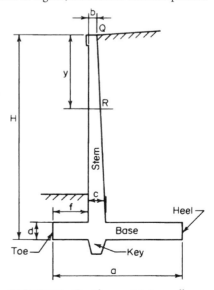

FIGURE 28. Cantilever retaining wall.

2.46 REINFORCED AND PRESTRESSED CONCRETE ENGINEERING AND DESIGN

active earth pressure; C_p = coefficient of passive earth pressure. Then

$$T = \frac{1}{2}C_a wy(y+2h) \qquad (51)$$

$$M = \left(\frac{1}{6}\right)C_a wy^2(y+3h) \qquad (52)$$

DESIGN OF A CANTILEVER RETAINING WALL

Applying the working-stress method, design a reinforced-concrete wall to retain an earth bank 14 ft (4.3 m) high. The top surface is horizontal and supports a surcharge of 500 lb/sq.ft. (23.9 kPa). The soil weighs 130 lb/ft³ (20.42 kN/m³), and its angle of internal friction is 35°; the coefficient of friction of soil and concrete is 0.5. The allowable soil pressure is 4000 lb/sq.ft. (191.5 kPa); f'_c = 3000 lb/sq.in. (20,685 kPa) and f_y = 40,000 lb/sq.in. (275,800 kPa). The base of the structure must be set 4 ft (1.2 m) below ground level to clear the frost line.

Calculation Procedure:

1. Secure a trial section of the wall

Apply these relations: $a = 0.60H$; $b \leq 8$ in. (203.2 mm); $c = d = b + 0.045h$; $f = a/3 - c/2$.

The trial section is shown in Fig. 29a, and the reinforcement is shown in Fig. 30. As the calculation will show, it is necessary to provide a key to develop the required resistance to sliding. The sides of the key are sloped to ensure that the surrounding soil will remain undisturbed during excavation.

2. Analyze the trial section for stability

The requirements are that there be a factor of safety (FS) against sliding and overturning of at least 1.5 and that the soil pressure have a value lying between 0 and 4000 lb/sq.ft. (0 and 191.5 kPa). Using the equation developed later in this handbook gives h = surcharge/soil weight = 500/130 = 3.85 ft (1.173 m); sin 35° = 0.574; tan 35° = 0.700; C_a = 0.271; C_p = 3.69; $C_a w$ = 35.2 lb/ft³ (5.53 kN/m³); $C_p w$ = 480 lb/ft³ (75.40 kN/m³); T_{AB} = ½(35.2)18(18 + 2 × 3.85) = 8140 lb (36,206.7 N); M_{AB} = (1/6)35.2(18)²(18 + 3 × 3.85) = 56,200 ft·lb (76,207.2 N·m).

The critical condition with respect to stability is that in which the surcharge extends to G. The moments of the stabilizing forces with respect to the toe are computed in Table 2. In Fig. 29c, x = 81,030/21,180 = 3.83 ft (1.167 m); e = 5.50 − 3.83 = 1.67 ft (0.509 m). The fact that the resultant strikes the base within the middle third attests to the absence of uplift. By $f = (P/A)(1 \pm 6e_x/d_x \pm 6e_y/d_y)$, p_a = (21,180/11)(1 + 6 × 1.67/11) = 3680 lb/sq.ft. (176.2 kPa); p_b = (21,180/11)(1 − 6 × 1.67/11) = 171 lb/sq.ft. (8.2 kPa). Check: x = (11/3)(3680 + 2 × 171)/(3680 + 171) = 3.83 ft (1.167 m), as before. Also, p_c = 2723 lb/sq.ft. (130.4 kPa); p_d = 2244 lb/sq.ft. (107.4 kPa); FS against overturning = 137,230/56,200 = 2.44. This is acceptable.

Lateral displacement of the wall produces sliding of earth on earth to the left of C and of concrete on earth to the right of C. In calculating the passive pressure, the layer of earth lying above the base is disregarded, since its effectiveness is unknown. The resistance to sliding is as follows: friction, A to C (Fig. 29): ½(3680 + 2723)(3)(0.700) = 6720 lb (29,890.6 N); friction, C to B: ½(2723 + 171)(8)(0.5) = 5790 lb (25,753.9 N); passive earth pressure:

TABLE 2. Stability of Retaining Wall

	Force, lb (N)			Arm, ft (m)	Moment, ft·lb (N·m)	
W_1 1.5(11)(150)	=	2,480	(11,031.0)	5.50 (1.676)	13,640	(18,495.8)
W_2 0.67(16.5)(150)	=	1,650	(7,339.2)	3.33 (1.015)	5,500	(7,458.0)
W_3 0.5(0.83)(16.5)(150)	=	1,030	(4,581.4)	3.95 (1.204)	4,070	(5,518.9)
W_4 1.25(1.13)(150)	=	210	(934.1)	3.75 (1.143)	790	(1,071.2)
W_5 0.5(0.83)(16.5)(130)	=	890	(3,958.7)	4.23 (1.289)	3,760	(5,098.6)
W_6 6.5(16.5)(130)	=	13,940	(62,005.1)	7.75 (2.362)	108,000	(146,448.0)
W_7 2.5(3)(130)	=	980	(4,359.1)	1.50 (0.457)	1,470	(1993.3)
Total		21,180	(94,208.6)		137,230	(186,083.8)
Overturning moment					56,200	(76,207.2)
Net moment about A					81,030	(109,876.6)

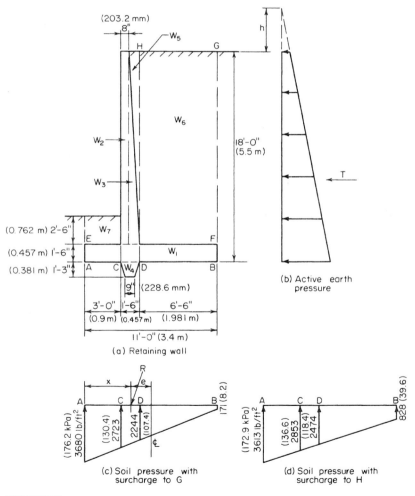

FIGURE 29

½(480)(2.75)² = 1820 lb (8095.4 N). The total resistance to sliding is the sum of these three items, or 14,330 lb (63,739.8 N). Thus, the FS against sliding is 14,330/8140 = 1.76. This is acceptable because it exceeds 1.5. Hence the trial section is adequate with respect to stability.

3. Calculate the soil pressures when the surcharge extends to H
Thus W_s = 500(6.5) = 3250 lb (14,456 N); ΣW = 21,180 + 3250 = 24,430 lb (108,664.6 N); M_a = 81,030 + 3250(7.75) = 106,220 ft·lb (144,034.3 N·m); x = 106,220/24,430 = 4.35 ft (1.326 m); e = 1.15 ft (0.351 m); p_a = 3613 lb/sq.ft. (173 kPa); p_b = 828 lb/sq.ft. (39.6 kPa); p_c = 2853 lb/sq.ft. (136.6 kPa); p_d = 2474 lb/sq.ft. (118.5 kPa).

4. Design the stem
At the base of the stem, y = 16.5 ft (5.03 m) and d = 18 − 3.5 = 14.5 in. (368.30 mm); T_{EF} = 7030 lb (31,269.4 N); M_{EF} = 538,000 in.·lb (60,783.24 N·m). The allowable shear at a distance d above the base is V_{allow} = vbd = 60(12)(14.5) = 10,440 lb (46,437.1 N). This is acceptable. Also, M_b = 223(12)(14.5)² = 563,000 in.·lb (63,607.74 N·m); therefore, the steel is stressed to capacity, and A_s = 538,000/[20,000(0.874)(14.5)] = 2.12 sq.in. (13.678 cm²). Use no. 9 bars 5½ in. (139.70 mm) on centers. Thus, A_s = 2.18 sq.in. (14.065 cm²); Σo = 7.7 in. (195.58/mm); u = 7030/[7.7(0.874)(14.5)] = 72 lb/sq.in. (496.5 kPa); u_{allow} = 235 lb/sq.in. (1620.3 kPa). This is acceptable.

Alternate bars will be discontinued at the point where they become superfluous. As the following calculations demonstrate, the theoretical cutoff point lies at y = 11 ft 7 in. (3.531 m), where M = 218,400 in.·lb (24,674.8 N·m); d = 4.5 + 10(11.58/16.5) = 11.52 in. (292.608 mm); A_s = 218,400/[20,000 (0.874)(11.52)] = 1.08 sq.in. (6.968 cm²). This is acceptable. Also, T = 3930 lb (17,480.6 N); u = 101 lb/sq.in. (696.4 kPa). This is acceptable. From the ACI *Code*, anchorage = 12(9/8) = 13.5 in. (342.9 mm).

The alternate bars will therefore be terminated at 6 ft 1 in. (1.854 m) above the top of the base. The *Code* requires that special precautions be taken where more than half the bars are spliced at a point of maximum stress. To circumvent this requirement, the short bars can be extended into the footing; therefore only the long bars require splicing. For the dowels, u_{allow} = 0.75(235) = 176 lb/sq.in. (1213.5 kPa); length of lap = 1.00(20,000)/[176(3.5)] = 33 in. (838.2 mm).

5. Design the heel
Let V and M denote the shear and bending moment, respectively, at section D. Case 1: surcharge extending to G—downward pressure p = 16.5(130) + 1.5(150) = 2370 lb/sq.ft. (113.5 kPa); V = 6.5[2370 − ½(2244 + 171)] = 7560 lb (33,626.9 N); M = 12(6.5)² [½ × 2370 − 1/6(2244 + 2 × 171)] = 383,000 in.·lb (43,271.3 N·m).

Case 2: surcharge extending to H—p = 2370 + 500 = 2870 lb/sq.ft. (137.4 kPa); V = 6.5[2870 − ½(2474 + 828)] = 7920 lb (35,228.1 N) < V_{allow}; M = 12(6.5)²[½ × 2870 − 1/6(2474 + 2 × 828)] = 379,000 in.·lb (42,819.4 N·m); A_s = 2.12(383/538) = 1.51 sq.in. (9.742 cm²).

To maintain uniform bar spacing throughout the member, use no. 8 bars 5½ in. (139.7 mm) on centers. In the heel, tension occurs at the top of the slab, and A_s = 1.72 sq.in. (11.097 cm²); Σo = 6.9 in. (175.26 mm); u = 91 lb/sq.in. (627.4 kPa); u_{allow} = 186 lb/sq.in. (1282.5 kPa). This is acceptable.

6. Design the toe
For this purpose, assume the absence of backfill on the toe, but disregard the minor modification in the soil pressure that results. Let V and M denote the shear and bending moment, respectively, at section C (Fig. 29). The downward pressure p = 1.5(150) = 225 lb/sq.ft. (10.8 kPa).

Case 1: surcharge extending to G (Fig. 29)—V = 3[½(3680 + 2723) − 225] = 8930 lb (39,720.6 N); M = 12(3)²[(1/6)(2723 + 2 × 3680) − ½(225)] = 169,300 in.·lb (19,127.5 N·m).

Case 2: surcharge extending to H (Fig. 29)—V = 9020 lb (40,121.0 N) < V_{allow}; M = 169,300 in.·lb (19,127.5 N·m); A_s = 2.12(169,300/538,000) = 0.67 sq.in. (4.323 cm²). Use no. 5 bars 5½ in. (139.7 mm) on centers. Then A_s = 0.68 sq.in. (4.387 cm²); Σo = 4.3 in. (109.22 mm); u = 166 lb/sq.in. (1144.4 kPa); u_{allow} = 422 lb/sq.in. (2909.7 kPa). This is acceptable.

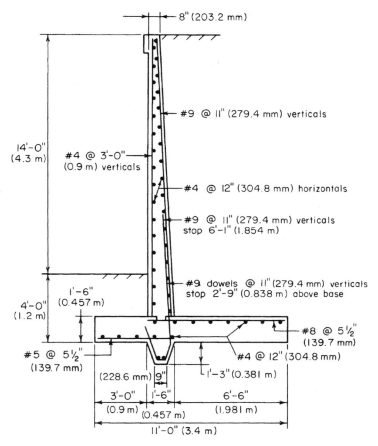

FIGURE 30

The stresses in the key are not amenable to precise evaluation. Reinforcement is achieved by extending the dowels and short bars into the key and bending them.

In addition to the foregoing reinforcement, no. 4 bars are supplied to act as temperature reinforcement and spacers for the main bars, as shown in Fig. 30.

PART 2
PRESTRESSED CONCRETE

Prestressed-concrete construction is designed to enhance the suitability of concrete as a structural material by inducing prestresses opposite in character to the stresses resulting from gravity loads. These prestresses are created by the use of steel wires or strands, called tendons, that are incorporated in the member and subjected to externally applied tensile forces. This prestressing of the steel may be performed either before or after pouring of the concrete. Thus, two methods of prestressing a concrete beam are available: pretensioning and posttensioning.

In pretensioning, the tendons are prestressed to the required amount by means of hydraulic jacks, their ends are tied to fixed abutments, and the concrete is poured around the tendons. When hardening of the concrete has advanced to the required state, the tendons are released. The tendons now tend to contract longitudinally to their original length and to expand laterally to their original diameter, both these tendencies being opposed by the surrounding concrete. As a result of the longitudinal restraint, the concrete exerts a tensile force on the steel and the steel exerts a compressive force on the concrete. As a result of the lateral restraint, the tendons are deformed to a wedge shape across a relatively short distance at each end of the member. It is within this distance, termed the *transmission length*, that the steel becomes bonded to the concrete and the two materials exert their prestressing forces on each other. However, unless greater precision is warranted, it is assumed for simplicity that the prestressing forces act at the end sections.

The tendons may be placed either in a straight line or in a series of straight-line segments, being deflected at designated points by means of holding devices. In the latter case, prestressing forces between steel and concrete occur both at the ends and at these deflection points.

In posttensioning, the procedure usually consists of encasing the tendons in metal or rubber hoses, placing these in the forms, and then pouring the concrete. When the concrete has hardened, the tendons are tensioned and anchored to the ends of the concrete beam by means of devices called *end anchorages*. If the hoses are to remain in the member, the void within the hose is filled with grout. Posttensioning has two important advantages compared with pretensioning: It may be performed at the job site, and it permits the use of parabolic tendons.

The term at transfer refers to the instant at which the prestressing forces between steel and concrete are developed. (In posttensioning, where the tendons are anchored to the concrete one at a time, in reality these forces are developed in steps.) Assume for simplicity that the tendons are straight and that the resultant prestressing force in these tendons lies below the centroidal axis of the concrete section. At transfer, the member cambers (deflects upward), remaining in contact with the casting bed only at the ends. Thus, the concrete beam is compelled to resist the prestressing force and to support its own weight simultaneously.

At transfer, the prestressing force in the steel diminishes because the concrete contracts under the imposed load. The prestressing force continues to diminish as time elapses as a result of the relaxation of the steel and the shrinkage and plastic flow of the concrete subsequent to transfer. To be effective, prestressed-concrete construction therefore requires the use of high-tensile steel in order that the reduction in prestressing force may be small in relation to the initial force. In all instances, we assume that the ratio of final to initial prestressing force is 0.85. Moreover, to simplify the stress calculations, we also assume that the full initial prestressing force exists at transfer and that the entire reduction in this force occurs during some finite interval following transfer.

Therefore, two loading states must be considered in the design: the initial state, in which the concrete sustains the initial prestressing force and the beam weight; and the final state, in which the concrete sustains the final prestressing force, the beam weight, and all superimposed loads. Consequently, the design of a prestressed-concrete beam differs from that of a conventional type in that designers must consider two stresses at each point, the initial stress and the final stress, and these must fall between the allowable compressive and tensile stresses. A beam is said to be in *balanced design* if the critical initial and final stresses coincide precisely with the allowable stresses.

The term *prestress* designates the stress induced by the *initial* prestressing force. The terms *prestress shear* and *prestress moment* refer to the vertical shear and bending moment, respectively, that the initial prestressing force induces in the concrete at a given section.

The *eccentricity* of the prestressing force is the distance from the action line of this resultant force to the centroidal axis of the section. Assume that the tendons are subjected to a uniform prestress. The locus of the centroid of the steel area is termed the *trajectory* of the steel or of the prestressing force.

The sign convention is as follows: The eccentricity is positive if the action line of the prestressing force lies below the centroidal axis. The trajectory has a positive slope if it inclines downward to the right. A load is positive if it acts downward. The vertical shear at a given section is positive if the portion of the beam to the left of this section exerts an upward force on the concrete. A bending moment is positive if it induces compression above the centroidal axis and tension below it. A compressive stress is positive; a tensile stress, negative.

The notational system is as follows. Cross-sectional properties: A = gross area of section, sq.in. (cm^2); A_s = area of prestressing steel, sq.in. (cm^2); d = effective depth of section at ultimate strength, in. (mm); h = total depth of section, in. (mm); I = moment of inertia of gross area, in^4 (cm^4); y_b = distance from centroidal axis to bottom fiber, in. (mm); S_b = section modulus with respect to bottom fiber = I/y_b, in^3 (cm^3); k_b = distance from centroidal axis to lower kern point, in. (mm); k_t = distance from centroidal axis to upper kern point, in. (mm). *Forces and moments*: F_i = initial prestressing force, lb (N); F_f = final prestressing force, lb (N); $\eta = F_f/F_i$; e = eccentricity of prestressing force, in. (mm); e_{con} = eccentricity of prestressing force having concordant trajectory; θ = angle between trajectory (or tangent to trajectory) and horizontal line; m = slope of trajectory; w = vertical load exerted by curved tendons on concrete in unit distance; w_w = unit beam weight; w_s = unit superimposed load; w_{DL} = unit dead load; w_{LL} = unit live load; w_u = unit ultimate load; V_p = prestress shear; M_p = prestress moment; M_w = bending moment due to beam weight; M_s = bending moment due to superimposed load; C_u = resultant compressive force at ultimate load; T_u = resultant tensile force at ultimate load. *Stresses*: f'_c = ultimate compressive strength of concrete, lb/sq.in. (kPa); f'_{ci} = compressive strength of concrete at transfer; f'_s = ultimate strength of prestressing steel; f_{su} = stress in prestressing steel at ultimate load; f_{bp} = stress in bottom fiber due to initial prestressing force; f_{bw} = bending stress in bottom fiber due to beam weight; f_{bs} = bending stress in bottom fiber due to superimposed loads; f_{bi} = stress in bottom fiber at initial state = $f_{bp} + f_{bw}$; f_{bf} = stress in bottom fiber at final state = $\eta f_{bp} + f_{bw} + f_{bs}$; f_{cai} = initial stress at centroidal axis. *Camber*: Δ_p = camber due to initial prestressing force, in. (mm); Δ_w = camber due to beam weight; Δ_i = camber at initial state; Δ_f = camber at final state.

The symbols that refer to the bottom fiber are transformed to their counterparts for the top fiber by replacing the subscript b with t. For example, f_{ti} denotes the stress in the top fiber at the initial state.

DETERMINATION OF PRESTRESS SHEAR AND MOMENT

The beam in Fig. 31a is simply supported at its ends and prestressed with an initial force of 300 kips (1334.4 kN). At section C, the eccentricity of this force is 8 in. (203.2 mm), and the slope of the trajectory is 0.014. (In the drawing, vertical distances are exaggerated in relation to horizontal distances.) Find the prestress shear and prestress moment at C.

2.52 REINFORCED AND PRESTRESSED CONCRETE ENGINEERING AND DESIGN

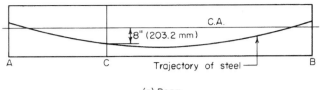

(a) Beam

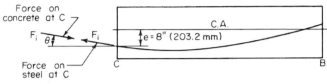

(b) Free-body diagram of CB

FIGURE 31

Calculation Procedure:

1. Analyze the prestressing forces
If the composite concrete-and-steel member is regarded as a unit, the prestressing forces that the steel exerts on the concrete are purely internal. Therefore, if a beam is simply supported, the prestressing force alone does not induce any reactions at the supports.

Refer to Fig. 31b, and consider the forces acting on the beam segment GB solely as a result of F_i. The left portion of the beam exerts a tensile force F_i on the tendons. Since GB is in equilibrium, the left portion also induces compressive stresses on the concrete at C, these stresses having a resultant that is numerically equal to and collinear with F_i.

2. Express the prestress shear and moment in terms of F_i
Using the sign convention described, express the prestress shear and moment in terms of F_i and θ. (The latter is positive if the slope of the trajectory is positive.) Thus $V_p = -F_i \sin \theta$; $M_p = -F_i e \cos \theta$.

3. Compute the prestress shear and moment
Since θ is minuscule, apply these approximations: $\sin \theta = \tan \theta$, and $\cos \theta = 1$. Then

$$V_p = -F_i \tan \theta \tag{53}$$

Or, $V_p = -300,000(0.014) = -4200$ lb ($-18,681.6$ N).
Also,

$$M_p = -F_i e \tag{54}$$

Or, $M_p = -300,000(8) = -2,400,000$ in.·lb ($-271,152$ N·m).

STRESSES IN A BEAM WITH STRAIGHT TENDONS

A 12 × 18 in. (304.8 × 457.2 mm) rectangular beam is subjected to an initial prestressing force of 230 kips (1023.0 kN) applied 3.3 in. (83.82 mm) below the center. The beam is on a simple span of 30 ft (9.1 m) and carries a superimposed load of 840 lb/lin ft (12,258.9 N/m).

Determine the initial and final stresses at the supports and at midspan. Construct diagrams to represent the initial and final stresses along the span.

Calculation Procedures:

1. Compute the beam properties
Thus, $A = 12(18) = 216$ sq.in. (1393.6 cm²); $S_b = S_t = (1/6)(12)(18)^2 = 648$ in³ (10,620.7 cm³); $w_w = (216/144)(150) = 225$ lb/lin ft (3,283.6 N/m).

2. Calculate the prestress in the top and bottom fibers
Since the section is rectangular, apply $f_{bp} = (F_i/A)(1 + 6e/h) = (230,000/216)(1 + 6 \times 3.3/18) = +2236$ lb/sq.in. (+15,417.2 kPa); $f_{tp} = (F_i/A)(1 - 6e/h) = -106$ lb/sq.in. (−730.9 kPa).
For convenience, record the stresses in Table 3 as they are obtained.

3. Determine the stresses at midspan due to gravity loads
Thus $M_s = (1/8)(840)(30)^2(12) = 1,134,000$ in.·lb (128,119.32 N·m); $f_{bs} = -1,134,000/648 = -1750$ lb/sq.in. (−12,066.3 kPa); $f_{ts} = +1750$ lb/sq.in. (12,066.3 kPa). By proportion, $f_{bw} = -1750(225/840) = -469$; $f_{tw} = +469$ lb/sq.in. (+3233.8 kPa).

4. Compute the initial and final stresses at the supports
Thus, $f_{bi} = +2236$ lb/sq.in. (+15,417.2 kPa); $f_{ti} = -106$ lb/sq.in. (−730.9 kPa); $f_{bf} = 0.85(2236) = +1901$ lb/sq.in. (+13,107.4 kPa); $f_{tf} = 0.85(-106) = -90$ lb/sq.in. (−620.6 kPa).

5. Determine the initial and final stresses at midspan
Thus $f_{bi} = +2236 - 469 = +1767$ lb/sq.in. (+12,183.5 kPa); $f_{ti} = -106 + 469 = +363$ lb/sq.in. (+2502.9 kPa); $f_{bf} = +1901 - 469 - 1750 = -318$ lb/sq.in. (−2192.6 kPa); $f_{tf} = -90 + 469 + 1750 = +2129$ lb/sq.in. (+14,679.5 kPa).

6. Construct the initial-stress diagram
In Fig. 32a, construct the initial-stress diagram A_tA_bBC at the support and the initial-stress diagram M_tM_bDE at midspan. Draw the parabolic arcs BD and CE. The stress diagram at

TABLE 3. Stresses in Prestressed-Concrete Beam

	At support		At midspan	
	Bottom fiber	Top fiber	Bottom fiber	Top fiber
(a) Initial prestress, lb/sq.in. (kPa)	+2,236 (+15,417.2)	−106 (−730.9)	+2,236 (+15,417.2)	−106 (−730.9)
(b) Final prestress, lb/sq.in. (kPa)	+1,901 (+13,107.4)	−90 (−620.6)	+1,901 (+13,107.4)	−90 (−620.6)
(c) Stress due to beam weight, lb/sq.in. (kPa)	.....	.....	−469 (−3,233.8)	+469 (3,233.8)
(d) Stress due to superimposed load, lb/sq.in. (kPa)	.....	.....	−1,750 (−12,066.3)	+1,750 (+12,066.3)
Initial stress: (a) + (c)	+2,236 (+15,417.2)	−106 (−730.9)	+1,767 (+12,183.5)	+363 (+2,502.9)
Final stress: (b) + (c) + (d)	+1,901 (+13,107.4)	−90 (−620.6)	−318 (−2,192.6)	+2,129 (+14,679.5)

2.54 REINFORCED AND PRESTRESSED CONCRETE ENGINEERING AND DESIGN

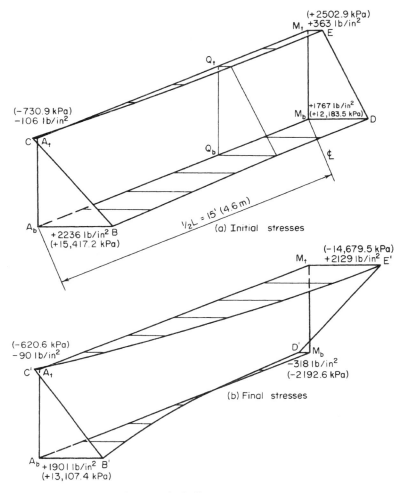

FIGURE 32. Isometric diagrams for half-span.

an intermediate section Q is obtained by passing a plane normal to the longitudinal axis. The offset from a reference line through B to the arc BD represents the value of f_{bw} at that section.

7. Construct the final-stress diagram
Construct Fig. 32b in an analogous manner. The offset from a reference line through B' to the arc $B'D'$ represents the value of $f_{bw} + f_{bs}$ at the given section.

8. Alternatively, construct composite stress diagrams for the top and bottom fibers
The diagram pertaining to the bottom fiber is shown in Fig. 33. The difference between the ordinates to DE and AB represents f_{bi} and the difference between the ordinates to FG and AC represents f_{bf}.

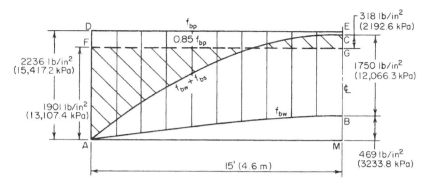

FIGURE 33. Stresses in bottom fiber along half-span.

This procedure illustrates the following principles relevant to a beam with straight tendons carrying a uniform load: At transfer, the critical stresses occur at the supports; under full design load, the critical stresses occur at midspan if the allowable final stresses exceed η times the allowable initial stresses in absolute value.

The primary objective in prestressed-concrete design is to maximize the capacity of a given beam by maximizing the absolute values of the prestresses at the section having the greatest superimposed-load stresses. The three procedures that follow, when taken as a unit, illustrate the manner in which the allowable prestresses may be increased numerically by taking advantage of the beam-weight stresses, which are opposite in character to the prestresses. The next procedure will also demonstrate that when a beam is not in balanced design, there is a range of values of F_i that will enable the member to carry this maximum allowable load. In summary, the objective is to maximize the capacity of a given beam and to provide the minimum prestressing force associated with this capacity.

DETERMINATION OF CAPACITY AND PRESTRESSING FORCE FOR A BEAM WITH STRAIGHT TENDONS

An 8×10 in. (203.2×254 mm) rectangular beam, simply supported on a 20-ft (6.1-m) span, is to be prestressed by means of straight tendons. The allowable stresses are: *initial*, $+2400$ and -190 lb/sq.in. ($+16,548$ and -1310.1 kPa); *final*, $+2250$ and -425 lb/sq.in. ($+15,513.8$ and -2930.3 kPa). Evaluate the allowable unit superimposed load, the maximum and minimum prestressing force associated with this load, and the corresponding eccentricities.

Calculation Procedure:

1. Compute the beam properties
Here $A = 80$ sq.in. (516.16 cm^2); $S = 133$ cu in (2179.9 cu cm); $w_w = 83$ lb/lin ft (1211.3 N/m).

2. Compute the stresses at midspan due to the beam weight
Thus, $M_w = (\frac{1}{8})(83)(20)^2(12) = 49,800$ in.·lb (5626.4 N·m); $f_{bw} = -49,800/133 = -374$ lb/sq.in. (-2578.7 kPa); $f_{tw} = +374$ lb/sq.in. (2578.7 kPa).

3. Set the critical stresses equal to their allowable values to secure the allowable unit superimposed load

Use Fig. 32 or 33 as a guide. At support: $f_{bi} = +2400$ lb/sq.in. (+16,548 kPa); $f_{ti} = -190$ lb/sq.in. (−1310.1 kPa); at midspan, $f_{bf} = 0.85(2400) - 374 + f_{bs} = -425$ lb/sq.in. (−2930.4 kPa); $f_{tf} = 0.85(-190) + 374 + f_{ts} = +2250$ lb/sq.in. (+15,513.8 kPa). Also, $f_{bs} = -2091$ lb/sq.in. (−14,417.4 kPa); $f_{ts} = +2038$ lb/sq.in. (+14,052 kPa).

Since the superimposed-load stresses at top and bottom will be numerically equal, the latter value governs the beam capacity. Or $w_s = w_w, f_{ts}/f_{tw} = 83(2038/374) = 452$ lb/lin ft (6596.4 N/m).

4. Find $F_{i,max}$ and its eccentricity

The value of w_s was found by setting the critical value of f_{ti} and of f_{tf} equal to their respective allowable values. However, since S_b is excessive for the load w_s, there is flexibility with respect to the stresses at the bottom. The designer may set the critical value of either f_{bi} or f_{bf} equal to its allowable value or produce some intermediate condition. As shown by the calculations in step 3, f_{bf} may vary within a range of 2091 − 2038 = 53 lb/sq.in. (365.4 kPa). Refer to Fig. 34, where the lines represent the stresses indicated.

Points B and F are fixed, but points A and E may be placed anywhere within the 53-lb/sq.in. (365.4-kPa) range. To maximize F_i, place A at its limiting position to the right; that is, set the critical value of f_{bi} rather than that of f_{bf} equal to the allowable value. Then $f_{cai} = F_{i,max}/A = \frac{1}{2}(2400 - 190) = +1105$ lb/sq.in. (+7619.0 kPa); $F_{i,max} = 1105(80) = 88,400$ lb (393,203.2 N); $f_{bp} = 1105 + 88,400e/133 = +2400$; $e = 1.95$ in. (49.53 mm).

5. Find $F_{i,min}$ and its eccentricity

For this purpose, place A at its limiting position to the left. Then $f_{bp} = 2,400 - (53/0.85) = +2338$ lb/sq.in. (+16,120.5 kPa); $f_{cai} = +1074$ lb/sq.in. (+7405.2 kPa); $F_{i,min} = 85,920$ lb (382,172.2 N); $e = 1.96$ in. (49.78 mm).

6. Verify the value of $F_{i,max}$ by checking the critical stresses

At support: $f_{bi} = +2400$ lb/sq.in. (+16,548.0 kPa); $f_{ti} = -190$ lb/sq.in. (−1310.1 kPa). At midspan: $f_{bf} = +2040 - 374 - 2038 = -372$ lb/sq.in. (−2564.9 kPa); $f_{tf} = -162 + 374 + 2038 = +2250$ lb/sq.in. (+15,513.8 kPa).

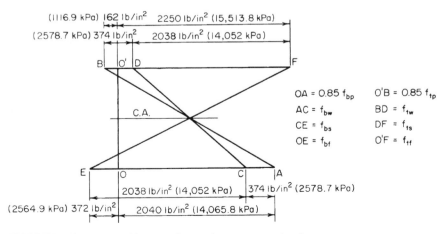

FIGURE 34. Stresses at midspan under maximum prestressing force.

7. Verify the value of $F_{i,min}$ by checking the critical stresses
At support: f_{bi} = +2338 lb/sq.in. (16,120.5 kPa); f_{ti} = −190 lb/sq.in. (−1310.1 kPa). At mid-span: f_{bf} = 0.85(2338) − 374 − 2038 = −425 lb/sq.in. (−2930.4 kPa); f_{tf} = +2250 lb/sq.in. (+15,513.8 kPa).

BEAM WITH DEFLECTED TENDONS

The beam in the previous calculation procedure is to be prestressed by means of tendons that are deflected at the quarter points of the span, as shown in Fig. 35a. Evaluate the allowable unit superimposed load, the magnitude of the prestressing force, the eccentricity e_1 in the center interval, and the maximum and minimum allowable values of the eccentricity e_2 at the supports. What increase in capacity has been obtained by deflecting the tendons?

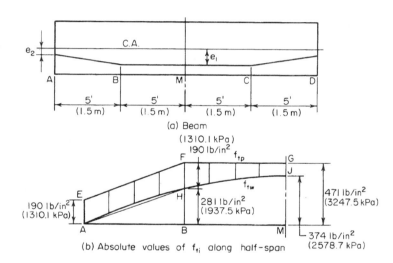

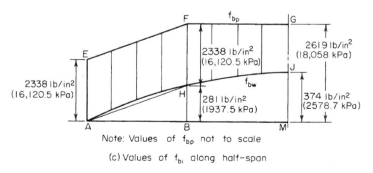

FIGURE 35

2.58 REINFORCED AND PRESTRESSED CONCRETE ENGINEERING AND DESIGN

Calculation Procedure:

1. Compute the beam-weight stresses at B
In the composite stress diagram, Fig. 35b, the difference between an ordinate to *EFG* and the corresponding ordinate to *AHJ* represents the value of f_{ti} at the given section. It is apparent that if *AE* does not exceed *HF*, then f_{ti} does not exceed *HE* in absolute value anywhere along the span. Therefore, for the center interval *BC*, the critical stresses at transfer occur at the boundary sections *B* and *C*. Analogous observations apply to Fig. 35c.
Computing the beam-weight stresses at *B* yields $f_{bw} = (¾)(-374) = -281$ lb/sq.in. (−1937.5 kPa); $f_{tw} = +281$ lb/sq.in. (+1937.5 kPa).

2. Tentatively set the critical stresses equal to their allowable values to secure the allowable unit superimposed load
Thus, at *B*: $f_{bi} = f_{bp} - 281 = +2400$; $f_{ti} = f_{tp} + 281 = -190$; $f_{bp} = +2681$ lb/sq.in. (+18,485.5 kPa); $f_{tp} = -471$ lb/sq.in. (−3247.5 kPa).
At *M*: $f_{bf} = 0.85(2681) - 374 + f_{bs} = -425$; $f_{tf} = 0.85(-471) + 374 + f_{ts} = +2250$; $f_{bs} = -2330$ lb/sq.in. (−16,065.4 kPa); $f_{ts} = +2277$ lb/sq.in. (+15,699.9 kPa). The latter value controls.
Also, $w_s = 83(2277/374) = 505$ lb/lin ft (7369.9 N/m); 505/452 = 1.12. The capacity is increased 12 percent.
When the foregoing calculations are compared with those in the previous calculation procedure, the effect of deflecting the tendons is to permit an increase of 281 lb/sq.in. (1937.5 kPa) in the absolute value of the prestress at top and bottom. The accompanying increase in f_{ts} is 0.85(281) = 239 lb/sq.in. (1647.9 kPa).

3. Find the minimum prestressing force and the eccentricity e_1
Examination of Fig. 34 shows that f_{cai} is not affected by the form of trajectory used. Therefore, as in the previous calculation procedure, $F_i = 85,920$ lb (382,172.2 N); $f_{tp} = 1074 - 85,920 e_1/133 = -471$; $e_1 = 2.39$ in. (60.706 mm).
Although it is not required, the value of $f_{bp} = 1074 + 1074 - (-471) = +2619$ lb/sq.in. (+18,058kPa), or $f_{bp} = 2681 - 53/0.85 = +2619$ lb/sq.in. (+18,058 kPa).

4. Establish the allowable range of values of e_2
At the supports, the tendons may be placed an equal distance above or below the center. Then $e_{2,max} = 1.96$ in. (23.44 mm); $e_{2,min} = -1.96$ in. (−23.44 mm).

DETERMINATION OF SECTION MODULI

A beam having a cross-sectional area of 500 sq.in. (3226 cm²) sustains a beam-weight moment equal to 3500 in.·kips (395.4 kN·m) at midspan and a superimposed moment that varies parabolically from 9000 in.·kips (1016.8 kN·m) at midspan to 0 at the supports. The allowable stresses are: *initial*, +2400 and −190 lb/sq.in. (+16,548 and −1310.1 kPa); *final*, +2250 and −200 lb/sq.in. (+15,513.8 and −1379 kPa). The member will be prestressed by tendons deflected at the quarter points. Determine the section moduli corresponding to balanced design, the magnitude of the prestressing force, and its eccentricity in the center interval. Assume that the calculated eccentricity is attainable (i.e., that the centroid of the tendons will fall within the confines of the section while satisfying insulation requirements).

Calculation Procedure:

1. Equate the critical initial stresses, and the critical final stresses, to their allowable values
Let M_w and M_s denote the indicated moments at midspan; the corresponding moments at the quarter point are three-fourths as large. The critical initial stresses occur at the quarter point,

while the critical final stresses occur at midspan. After equating the stresses to their allowable values, solve the resulting simultaneous equations to find the section moduli and prestresses. Thus: *stresses in bottom fiber,* $f_{bi} = f_{bp} - 0.75M_w/S_b = +2400$; $f_{bf} = 0.85f_{bp} - M_w/S_b - M_s/S_b = -200$. Solving gives $S_b = (M_s + 0.3625M_w)/2240 = 4584$ in^3 (75,131.7 cm^3) and $f_{bp} = +2973$ lb/sq.in. (+20,498.8 kPa); *stresses in top fiber,* $f_{ti} = f_{tp} + 0.75(M_w/S_t) = -190$; $f_{tf} = 0.85f_{tp} + M_w/S_t + M_s/S_t = +2250$. Solving yields $S_t = (M_s + 0.3625M_w)/2412 = 4257$ in^3 (69,772.2 cm^3) and $f_{tp} = -807$ lb/sq.in. (−5564.2 kPa).

2. Evaluate F_i and e
In this instance, e denotes the eccentricity in the center interval. Thus $f_{bp} = F_i/A + F_i e/S_b = +2973$; $f_{tp} = F_i/A - F_i e/S_t = -807$; $F_i = (2973S_b - 807S_t)A/(S_b + S_t) = 576,500$ lb (2,564,272.0 N); $e = 2973S_b/F_i - S_b/A = 14.47$ in. (367.538 mm).

PRESTRESSED-CONCRETE BEAM DESIGN GUIDES

On the basis of the previous calculation procedures, what conclusions may be drawn that will serve as guides in the design of prestressed-concrete beams?

Calculation Procedure:

1. Evaluate the results obtained with different forms of tendons
The capacity of a given member is increased by using deflected rather than straight tendons, and the capacity is maximized by using parabolic tendons. (However, in the case of a pretensioned beam, an economy analysis must also take into account the expense incurred in deflecting the tendons.)

2. Evaluate the prestressing force
For a given ratio of y_b/y_t, the prestressing force that is required to maximize the capacity of a member is a function of the cross-sectional area and the allowable stresses. It is independent of the form of the trajectory.

3. Determine the effect of section moduli
If the section moduli are in excess of the minimum required, the prestressing force is minimized by setting the critical values of f_{bf} and f_{ti} equal to their respective allowable values. In this manner, points A and B in Fig. 34 are placed at their limiting positions to the left.

4. Determine the most economical short-span section
For a short-span member, an I section is most economical because it yields the required section moduli with the minimum area. Moreover, since the required values of S_b and S_t differ, the area should be disposed unsymmetrically about middepth to secure these values.

5. Consider the calculated value of e
Since an increase in span causes a greater increase in the theoretical eccentricity than in the depth, the calculated value of e is not attainable in a long-span member because the centroid of the tendons would fall beyond the confines of the section. For this reason, long-span members are generally constructed as T sections. The extensive flange area elevates the centroidal axis, thus making it possible to secure a reasonably large eccentricity.

6. Evaluate the effect of overload
A relatively small overload induces a disproportionately large increase in the tensile stress in the beam and thus introduces the danger of cracking. Moreover, owing to the presence of many variable quantities, there is not a set relationship between the beam capacity at allowable final stress and the capacity at incipient cracking. It is therefore imperative that every prestressed-concrete beam be subjected to an ultimate-strength

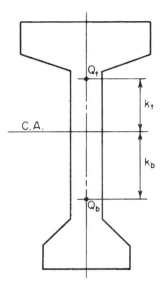

FIGURE 36. Kern points.

analysis to ensure that the beam provides an adequate factor of safety.

KERN DISTANCES

The beam in Fig. 36 has the following properties: $A = 850$ sq.in. (5484.2 cm^2); $S_b = 11,400$ in^3 (186,846.0 cm^3); $S_t = 14,400$ in^3 (236,016.0 cm^3). A prestressing force of 630 kips (2802.2 kN) is applied with an eccentricity of 24 in. (609.6 mm) at the section under investigation. Calculate f_{bp} and f_{tp} by expressing these stresses as functions of the kern distances of the section.

Calculation Procedure:

1. Consider the prestressing force to be applied at each kern point, and evaluate the kern distances

Let Q_b and Q_t denote the points at which a compressive force must be applied to induce a zero stress in the top and bottom fiber, respectively. These are referred to as the *kern points* of the section, and the distances k_b and k_t, from the centroidal axis to these points are called the *kern distances*.

Consider the prestressing force to be applied at each kern point in turn. Set the stresses f_{tp} and f_{bp} equal to zero to evaluate the kern distances k_b and k_t, respectively. Thus $f_{tp} = F_i/A - F_i k_b/S_t = 0$, Eq. a; $f_{bp} = F_i/A - F_i k_t/S_b = 0$, Eq. b. Then

$$k_b = \frac{S_t}{a} \quad \text{and} \quad k_t = \frac{S_b}{A} \tag{55}$$

And, $k_b = 14,400/850 = 16.9$ in. (429.26 mm); $k_t = 11,400/850 = 13.4$ in. (340.36 mm).

2. Express the stresses f_{bp} and f_{tp} associated with the actual eccentricity as functions of the kern distances

By combining the stress equations with Eqs. a and b, the following equations are obtained:

$$f_{bp} = \frac{F_i(k_t + e)}{S_b} \quad \text{and} \quad f_{tp} = \frac{F_i(k_b + e)}{S_t} \tag{56}$$

Substituting numerical values gives $f_{bp} = 630,000(13.4 + 24)/11,400 = +2067$ lb/sq.in. (+14,252.0 kPa); $f_{tp} = 630,000(16.9 - 24)14,400 = -311$ lb/sq.in. (−2144.3 kPa).

3. Alternatively, derive Eq. 56 by considering the increase in prestress caused by an increase in eccentricity

Thus, $\Delta f_{bp} = F_i \Delta e/S_b$; therefore, $f_{bp} = F_i(k_t + e)/S_b$.

MAGNEL DIAGRAM CONSTRUCTION

The data pertaining to a girder having curved tendons are $A = 500$ sq.in. (3226.0 cm^2); $S_b = 5000$ in^3 (81,950 cm^3); $S_t = 5340$ in^3 (87,522.6 cm^3); $M_w = 3600$ in.·kips (406.7 kN·m); $M_s = 9500$ in.·kips (1073.3 kN·m). The allowable stresses are: *initial*, +2400 and −190 lb/sq.in. (+16,548 and −1310.1 kPa); *final*, +2250 and −425 lb/sq.in. (+15,513.8 and

−2930.4 kPa). (*a*) Construct the Magnel diagram for this member. (*b*) Determine the minimum prestressing force and its eccentricity by referring to the diagram. (*c*) Determine the prestressing force if the eccentricity is restricted to 18 in. (457.2 mm).

Calculation Procedure:

1. Set the initial stress in the bottom fiber at midspan equal to or less than its allowable value, and solve for the reciprocal of F_i

In this situation, the superimposed load is given, and the sole objective is to minimize the prestressing force. The Magnel diagram is extremely useful for this purpose because it brings into sharp focus the relationship between F_i and e. In this procedure, let f_{bi} and f_{bf} and so forth represent the allowable stresses.
Thus,

$$\frac{1}{F_i} \geq \frac{k_t + e}{M_w + f_{bt}S_b} \tag{57a}$$

2. Set the final stress in the bottom fiber at midspan equal to or algebraically greater than its allowable value, and solve for the reciprocal of F_i
Thus,

$$\frac{1}{F_i} \leq \frac{\eta(k_t + e)}{M_w + M_s + f_{bf}S_b} \tag{57b}$$

3. Repeat the foregoing procedure with respect to the top fiber
Thus,

$$\frac{1}{F_i} \geq \frac{e - k_b}{M_w + f_{ti}S_t} \tag{57c}$$

and

$$\frac{1}{F_i} \leq \frac{\eta(e - k_b)}{M_w + M_s + f_{tf}S_b} \tag{57d}$$

4. Substitute numerical values, expressing F_i in thousands of kips
Thus, $1/F_i \geq (10 + e)/15.60$, Eq. *a*; $1/F_i$ $(10 + e)/12.91$, Eq. *b*; $1/F_i \leq (e - 10.68)/4.61$, Eq. *c*; $1/F_i \leq (e - 10.68)/1.28$, Eq. *d*.

5. Construct the Magnel diagram
In Fig. 37, consider the foregoing relationships as equalities, and plot the straight lines that represent them. Each point on these lines represents a set of values of $1/F_i$ and e at which the designated stress equals its allowable value.

When the section moduli are in excess of those corresponding to balanced design, as they are in the present instance, line *b* makes a greater angle with the *e* axis than does *a*, and line *d* makes a greater angle than does *c*. From the sense of each inequality, it follows that $1/F_i$ and *e* may have any set of values represented by a point within the quadrilateral *CDEF* or on its circumference.

6. To minimize F_i, determine the coordinates of point E at the intersection of lines b and c

Thus, $1/F_i = (10 + e)/12.91 = (e - 10.68)/4.61$; so $e = 22.2$ in. (563.88 mm); $F_i = 401$ kips (1783.6 kN).

The Magnel diagram confirms the third design guide presented earlier in the section.

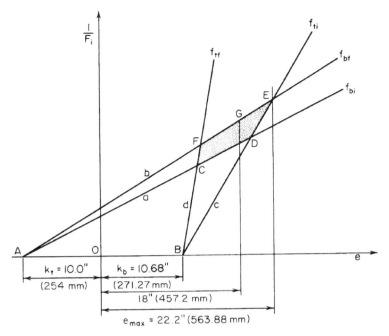

FIGURE 37. Magnel diagram.

7. For the case where e is restricted to 18 in. (457.2 mm), minimize F_i by determining the ordinate of point G on line b

Thus, in Fig. 37, $1/F_i = (10 + 18)/12.91$; $F_i = 461$ kips (2050.5 kN).

The Magnel diagram may be applied to a beam having deflected tendons by substituting for M_w in Eqs. 57a and 57c the beam-weight moment at the deflection point.

CAMBER OF A BEAM AT TRANSFER

The following pertain to a simply supported prismatic beam: $L = 36$ ft (11.0 m); $I = 40,000$ in^4 (166.49 dm^4); $f'_{ci} = 4000$ lb/sq.in. (27,580.0 kPa); $w_w = 340$ lb/lin ft (4961.9 N/m); $F_i = 430$ kips (1912.6 kN); $e = 8.8$ in. (223.5 mm) at midspan. Calculate the camber of the member at transfer under each of these conditions: (a) the tendons are straight across the entire span; (b) the tendons are deflected at the third points, and the eccentricity at the supports is zero; (c) the tendons are curved parabolically, and the eccentricity at the supports is zero.

Calculation Procedure:

1. Evaluate E_c at transfer, using the ACI Code

Review the moment-area method of calculating beam deflections, which is summarized earlier. Consider an upward displacement (camber) as positive, and let the symbols Δ_p, Δ_w, and Δ_i, defined earlier, refer to the camber at midspan.

Thus, using the ACI Code, $E_c = (145)^{1.5}(33)(4000)^{0.5} = 3,644,000$ lb/sq.in. (25,125.4 MPa).

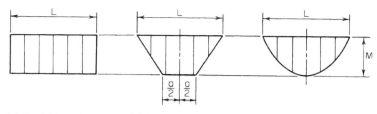

(a) Straight tendons   (b) Deflected tendons   (c) Parabolic tendons

FIGURE 38. Prestress-moment diagrams.

2. Construct the prestress-moment diagrams associated with the three cases described

See Fig. 38. By symmetry, the elastic curve corresponding to F_i is horizontal at midspan. Consequently, Δ_p equals the deviation of the elastic curve at the support from the tangent to this curve at midspan.

3. Using the literal values shown in Fig. 38, develop an equation for Δ_p by evaluating the tangential deviation; substitute numerical values

Thus, case a:

$$\Delta_p = \frac{ML^2}{8E_c I} \tag{58}$$

or $\Delta_p = 430{,}000(8.8)(36)^2(144)/[8(3{,}644{,}000)(40{,}000)] = 0.61$ in. (15.494 mm). For case b:

$$\Delta_p = \frac{M(2L^2 + 2La - a^2)}{24 E_c I} \tag{59}$$

or $\Delta_p = 0.52$ in. (13.208 mm). For case c:

$$\Delta_p = \frac{5 ML^2}{48 E_c I} \tag{60}$$

or $\Delta_p = 0.51$ in. (12.954 mm).

4. Compute Δ_w

Thus, $\Delta_w = -5 w_w L^4/(384 E_c I) = -0.09$ in. (−2.286 mm).

5. Combine the foregoing results to obtain Δ_i

Thus: case a, $\Delta_i = 0.61 - 0.09 = 0.52$ in. (13.208 mm); case b, $\Delta_i = 0.52 - 0.09 = 0.43$ in. (10.922 mm); case c, $\Delta_i = 0.51 - 0.09 = 0.42$ in. (10.688 mm).

DESIGN OF A DOUBLE-T ROOF BEAM

The beam in Fig. 39 was selected for use on a simple span of 40 ft (12.2 m) to carry the following loads: roofing, 12 lb/sq.ft. (574.5 N/m²); snow, 40 lb/sq.ft. (1915.1 N/m²); total, 52 lb/sq.ft. (2489.6 N/m²). The member will be pretensioned with straight seven-wire strands, $7/16$ in. (11.11 mm) diameter, having an area of 0.1089 sq.in. (0.70262 cm²) each

2.64 REINFORCED AND PRESTRESSED CONCRETE ENGINEERING AND DESIGN

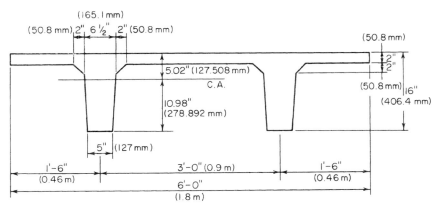

FIGURE 39. Double-T roof beam.

and an ultimate strength of 248,000 lb/sq.in. (1,709,960.0 kPa). The concrete strengths are $f'_c = 5000$ lb/sq.in. (34,475.0 kPa) and $f'_{ci} = 4000$ lb/sq.in. (27,580.0 kPa). The allowable stresses are: *initial*, +2400 and −190 lb/sq.in. (+16,548.0 and −1310.1 kPa); *final*, +2250 and −425 lb/sq.in. (+15,513.8 and −2930.4 kPa). Investigate the adequacy of this section, and design the tendons. Compute the camber of the beam after the concrete has hardened and all dead loads are present. For this calculation, assume that the final value of E_c is one-third of that at transfer.

Calculation Procedure:

1. Compute the properties of the cross section
Let f_{bf} and f_{tf} denote the respective stresses at *midspan* and f_{bi} and f_{ti} denote the respective stresses *at the support*. Previous calculation procedures demonstrated that where the section moduli are excessive, the minimum prestressing force is obtained by setting f_{bf} and f_{ti} equal to their allowable values.
Thus $A = 316$ sq.in. (2038.8 cm²); $I = 7240$ in⁴ (30.14 dm⁴); $y_b = 10.98$ in. (278.892 mm); $y_t = 5.02$ in. (127.508 mm); $S_b = 659$ in³ (10,801.0 cm³); $S_t = 1442$ in³ (23,614 cm³); $w_w = (316/144)150 = 329$ lb/lin ft (4801.4 N/m).

2. Calculate the total midspan moment due to gravity loads and the corresponding stresses
Thus $w_s = 52(6) = 312$ lb/lin ft (4553.3 N/m); $w_w = 329$ lb/lin ft (4801.4 N/m); and $M_w + M_s = (\frac{1}{8})(641)(40^2)(12) = 1{,}538{,}000$ in.·lb (173,763.2 N·m); $f_{bw} + f_{bs} = -1{,}538{,}000/659 = -2334$ lb/sq.in. (−16,092.9 kPa); $f_{tw} + f_{ts} = +1{,}538{,}000/1442 = +1067$ lb/sq.in. (+7357.0 kPa).

3. Determine whether the section moduli are excessive
Do this by setting f_{bf} and f_{ti} equal to their allowable values and computing the corresponding values of f_{bi} and f_{tf}. Thus, $f_{bf} = 0.85 f_{bp} − 2334 = −425$; therefore, $f_{bp} = +2246$ lb/sq.in. (+15,486.2 kPa); $f_{ti} = f_{tp} = −190$ lb/sq.in. (−1310.1 kPa); $f_{bi} = f_{bp} = +2246 < 2400$ lb/sq.in. (+16,548.0 kPa). This is acceptable. Also, $f_{tf} = 0.85(−190) + 1067 = +905 < 2250$ lb/sq.in. (+15,513.8 kPa); this is acceptable. The section moduli are therefore excessive.

4. Find the minimum prestressing force and its eccentricity
Refer to Fig. 40. Thus, $f_{bp} = +2246$ lb/sq.in. (+15,486.2 kPa); $f_{tp} = −190$ lb/sq.in. (−1310.1 kPa); slope of $AB = 2246 − (−190)/16 = 152.3$ lb/(sq.in.·in.) (41.33 MPa/m);

$F_i/A = CD = 2246 - 10.98\ (152.3) = 574$ lb/sq.in. (3957.7 kPa); $F_i = 574(316) = 181{,}400$ lb (806,867.2 N); slope of $AB = F_i e/I = 152.3$; $e = 152.3(7240)/181{,}400 = 6.07$ in. (154.178 mm).

5. Determine the number of strands required, and establish their disposition

In accordance with the ACI *Code*, allowable initial force per strand = $0.1089\ (0.70)(248{,}000) = 18{,}900$ lb (84,067.2 N); number required = $181{,}400/18{,}900 = 9.6$. Therefore, use 10 strands (5 in each web) stressed to 18,140 lb (80,686.7 N) each.

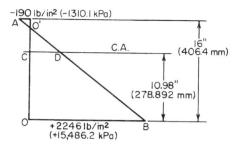

FIGURE 40. Prestress diagram.

Referring to the ACI *Code* for the minimum clear distance between the strands, we find the allowable center-to-center spacing = $4(^7/_{16}) = 1\tfrac{3}{4}$ in. (44.45 mm). Use a 2-in. (50.8-mm) spacing. In Fig. 41, locate the centroid of the steel, or $y = (2 \times 2 + 1 \times 4)/5 = 1.60$ in. (40.64 mm); $v = 10.98 - 6.07 - 1.60 = 3.31$ in. (84.074 mm); set $v = 3^5/_{16}$ in. (84.138 mm).

6. Calculate the allowable ultimate moment of the member in accordance with the ACI Code

Thus, $A_s = 10(0.1089) = 1.089$ sq.in. (7.0262 cm^2); $d = y_t + e = 5.02 + 6.07 = 11.09$ in. (281.686 mm); $p = A_s/(bd) = 1.089/[72(11.09)] = 0.00137$.

Compute the steel stress and resultant tensile force at ultimate load:

$$f_{su} = f'_s\left(1 - \frac{0.5 p f'_s}{f'_c}\right) \qquad (61)$$

Or, $f_{su} = 248{,}000(1 - 0.5 \times 0.00137 \times 248{,}000/5000) = 240{,}000$ lb/sq.in. (1,654,800 kPa); $T_u = A_s f_{su} = 1.089(240{,}000) = 261{,}400$ lb (1,162,707.2 N).

Compute the depth of the compression block. This depth, a, is found from $C_u = 0.85(5000)(72a) = 261{,}400$ lb (1,162,707.2 N); $a = 0.854$ in. (21.6916 mm); $jd = d - a/2 = 10.66$ in. (270.764 mm); $M_u = \phi T_u jd = 0.90(261{,}400)(10.66) = 2{,}500{,}000$ in.·lb (282,450.0 N·m).

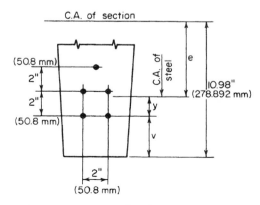

FIGURE 41. Location of tendons.

Calculate the steel index to ascertain that it is below the limit imposed by the ACI *Code*, or $q = pf_{su}/f_c' = 0.00137\,(240{,}000)/5000 = 0.0658 < 0.30$. This is acceptable.

7. Calculate the required ultimate-moment capacity as given by the ACI Code

Thus, $w_{DL} = 329 + 12(6) = 401$ lb/lin ft (5852.2 N/m); $w_{LL} = 40(6) = 240$ lb/lin ft (3502.5 N/m); $w_u = 1.5w_{DL} + 1.8w_{LL} = 1034$ lb/lin ft (15,090.1 N/m); M_u required $= (\tfrac{1}{8})(1034)(40)^2(12) = 2{,}480{,}000 < 2{,}500{,}000$ in.·lb (282,450.0 N·m). The member is therefore adequate with respect to its ultimate-moment capacity.

8. Calculate the maximum and minimum area of web reinforcement in the manner prescribed in the ACI Code

Since the maximum shearing stress does not vary linearly with the applied load, the shear analysis is performed at ultimate-load conditions. Let A_v = area of web reinforcement placed perpendicular to the longitudinal axis; V_c' = ultimate-shear capacity of concrete; V_p' = vertical component of F_f at the given section; V_u' = ultimate shear at given section; s = center-to-center spacing of stirrups; f_{pc}' = stress due to F_f, evaluated at the centroidal axis, or at the junction of the web and flange when the centroidal axis lies in the flange.

Calculate the ultimate shear at the critical section, which lies at a distance $d/2$ from the face of the support. Then distance from midspan to the critical section $= \tfrac{1}{2}(L - d) = 19.54$ ft (5.955 m); $V_u' = 1034(19.54) = 20{,}200$ lb (89,849.6 N).

Evaluate V_c' by solving the following equations and selecting the smaller value:

$$V_{ci}' = 1.7 b' d (f_c')^{0.5} \tag{62}$$

where d = effective depth, in. (mm); b' = width of web at centroidal axis, in. (mm); $b' = 2(5 + 1.5 \times 10.98/12) = 12.74$ in. (323.596 mm); $V_{ci}' = 1.7(12.74)(11.09)(5000)^{0.5} = 17{,}000$ lb (75,616.0 N). Also,

$$V_{cw}' = b'd(3.5 f_c'^{\,0.5} + 0.3 f_{pc}') + V_p' \tag{63}$$

where d = effective depth or 80 percent of the overall depth, whichever is greater, in. (mm). Thus, $d = 0.80(16) = 12.8$ in. (325.12 mm); $V_p' = 0$. From step 4, $f_{pc}' = 0.85(574) = +488$ lb/sq.in. (3364.8 kPa); $V_{cw}' = 12.74(12.8)(3.5 \times 5000^{0.5} + 0.3 \times 488) = 64{,}300$ lb (286,006.4 N); therefore, $V_c' = 17{,}000$ lb (75,616.0 N).

Calculate the maximum web-reinforcement area by applying the following equation:

$$A_v = \frac{S(V_u' - \phi V_c')}{\phi d f_y} \tag{64}$$

where d = effective depth at section of maximum moment, in. (mm). Use $f_y = 40{,}000$ lb/sq.in. (275,800.0 kPa), and set $s = 12$ in. (304.8 mm). Then $A_v = 12(20{,}200 - 0.85 \times 17{,}000)/[0.85(11.09)(40{,}000)] = 0.184$ sq.in./ft (3.8949 cm²/m). This is the area required at the ends.

Calculate the minimum web-reinforcement area by applying

$$A_v = \frac{A_s}{80} \frac{f_s'}{f_y} \frac{s}{(b'd)^{0.5}} \tag{65}$$

or $A_v = (1.089/80)(248{,}000/40{,}000)12/(12.74 \times 11.09)^{0.5} = 0.085$ sq.in./ft (1.7993 cm²/m).

9. Calculate the camber under full dead load

From the previous procedure, $E_c = (\frac{1}{3})(3.644)(10)^6 = 1.215 \times 10^6$ lb/sq.in. $(8.377 \times 10^6$ kPa); $E_c I = 1.215(10)^6(7240) = 8.8 \times 10^9$ lb·sq.in. $(25.25 \times 10^6$ N·m²); $\Delta_{ADL} = -5(401)(40)^4(1728)/[384(8.8)(10)^9] = -2.62$ in. $(-66.548$ mm). By Eq. 58, $\Delta_p = 0.85(181,400)(6.07)(40)^2(144)/[8(8.8)(10)^9] = 3.06$ in. $(77.724$ mm); $\Delta = 3.06 - 2.62 = 0.44$ in. $(11.176$ mm).

DESIGN OF A POSTTENSIONED GIRDER

The girder in Fig. 42 has been selected for use on a 90-ft (27.4-m) simple span to carry the following superimposed loads: dead load, 1160 lb/lin ft (16,928.9 N/m); live load, 1000 lb/lin ft (14,593.9 N/m). The girder will be posttensioned with Freyssinet cables. The concrete strengths are $f'_c = 5000$ lb/sq.in. (34,475 kPa) and $f'_{ci} = 4000$ lb/sq.in. (27,580 kPa). The allowable stresses are: *initial*, +2400 and −190 lb/sq.in. (+16,548 and −1310.1 kPa); *final*, +2250 and −425 lb/sq.in. (+15,513.8 and −2930.4 kPa). Complete the design of this member, and calculate the camber at transfer.

Calculation Procedure:

1. Compute the properties of the cross section

Since the tendons will be curved, the initial stresses at midspan may be equated to the allowable values. The properties of the cross section are $A = 856$ sq.in. (5522.9 cm²); $I = 394,800$ in⁴ (1643 dm⁴); $y_b = 34.6$ in. (878.84 mm); $y_t = 27.4$ in. (695.96 mm); $S_b = 11,410$ in³ (187,010 cm³); $S_t = 14,410$ in³ (236,180 cm³); $w_w = 892$ lb/lin ft (13,017.8 N/m).

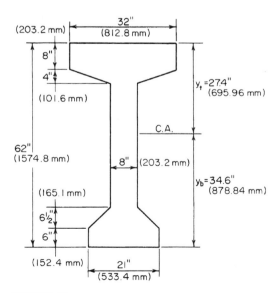

FIGURE 42

2. Calculate the stresses at midspan caused by gravity loads

Thus $f_{bw} = -950$ lb/sq.in. (-6550.3 kPa); $f_{bs} = -2300$ lb/sq.in. ($-15,858.5$ kPa); $f_{tw} = +752$ lb/sq.in. ($+5185.0$ kPa); $f_{ts} = +1820$ lb/sq.in. ($+12,548.9$ kPa).

3. Test the section adequacy

To do this, equate f_{bf} and f_{ti} to their allowable values and compute the corresponding values of f_{bi} and f_{tf}. Thus $f_{bf} = 0.85 f_{bp} - 950 - 2300 = -425$; $f_{ti} = f_{tp} + 752 = -190$; therefore, $f_{bp} = +3324$ lb/sq.in. ($+22,919.0$ kPa) and $f_{tp} = -942$ lb/sq.in. (-6495.1 kPa); $f_{bi} = +3324 - 950 = +2374 < 2400$ lb/sq.in. ($16,548.0$ kPa). This is acceptable. And $f_{tf} = 0.85(-942) + 752 + 1820 = +1771 < 2250$ lb/sq.in. ($15,513.8$ kPa). This is acceptable. The section is therefore adequate.

4. Find the minimum prestressing force and its eccentricity at midspan

Do this by applying the prestresses found in step 3. Refer to Fig. 43. Slope of $AB = [3324 - (-942)]/62 = 68.8$ lb/(sq.in.·in) (18.68 kPa/mm); $F_i/A = CD = 3324 - 34.6(68.8) = 944$ lb/sq.in. (6508.9 kPa); $F_i = 944(856) = 808,100$ lb ($3,594,428.8$ N); slope of $AB = F_i e/I = 68.8$; $e = 68.8(394,800)/808,100 = 33.6$ in. (853.44 mm). Since $y_b = 34.6$ in. (878.84 mm), this eccentricity is excessive.

5. Select the maximum feasible eccentricity; determine the minimum prestressing force associated with this value

Try $e = 34.6 - 3.0 = 31.6$ in. (802.64 mm). To obtain the minimum value of F_i, equate f_{bf} to its allowable value. Check the remaining stresses. As before, $f_{bp} = +3324$ lb/sq.in. ($+22,919$ kPa). But $f_{bp}, = F_i/856 + 31.6F_i/11,410 = +3324$; therefore $F_i = 844,000$ lb (3754.1 kN). Also, $f_{tp} = -865$ lb/sq.in. (-5964.2 kPa); $f_{bi} = +2374$ lb/sq.in. ($+16,368.7$ kPa); $f_{ti} = -113$ lb/sq.in. (-779.1 kPa); $f_{tf} = +1837$ lb/sq.in. ($+12,666.1$ kPa).

6. Design the tendons, and establish their pattern at midspan

Refer to a table of the properties of Freyssinet cables, and select 12/0.276 cables. The designation indicates that each cable consists of 12 wires of 0.276-in. (7.0104-mm) diameter. The ultimate strength is 236,000 lb/sq.in. (1,627,220 kPa). Then $A_s = 0.723$ sq.in. (4.6648 cm^2) per cable. Outside diameter of cable = $1^{5}/_{8}$ in. (41.27 mm). Recommended final prestress = 93,000 lb (413,664 N) per cable; initial prestress = 93,000/0.85 = 109,400 lb (486,611.2 N) per cable. Therefore, use eight cables at an initial prestress of 105,500 lb (469,264.0 N) each.

A section of the ACI *Code* requires a minimum cover of 1½ in. (38.1 mm) and another section permits the ducts to be bundled at the center. Try the tendon pattern shown in Fig. 44. Thus, $y = [6(2.5) + 2(4.5)]/8 = 3.0$ in. (76.2 mm). This is acceptable.

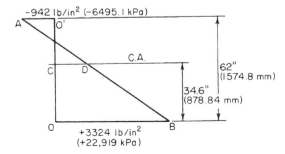

FIGURE 43. Prestress diagram.

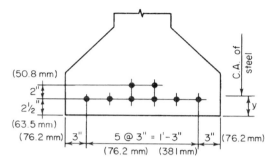

FIGURE 44. Location of tendons at midspan.

7. Establish the trajectory of the prestressing force
Construct stress diagrams to represent the initial and final stresses in the bottom and top fibers along the entire span.

For convenience, set $e = 0$ at the supports. The prestress at the ends is therefore $f_{bp} = f_{tp} = 844,000/856 = +986$ lb/sq.in. (+6798.5 kPa). Since e varies parabolically from maximum at midspan to zero at the supports, it follows that the prestresses also vary parabolically.

In Fig. 45a, draw the parabolic arc AB with summit at B to represent the absolute value of f_{bp}. Draw the parabolic arc OC in the position shown to represent f_{bw}. The vertical distance between the arcs at a given section represents the value of f_{bi}; this value is maximum at midspan.

In Fig. 45b, draw $A'B'$ to represent the absolute value of the final prestress; draw OC' to represent the absolute value of $f_{bw} + f_{bs}$. The vertical distance between the arcs represents the value of f_{bf}. This stress is compressive in the interval ON and tensile in the interval NM.

Construct Fig. 45c and d in an analogous manner. The stress f_{ti} is compressive in the interval OQ.

8. Calculate the allowable ultimate moment of the member
The midspan section is critical in this respect. Thus, $d = 62 - 3 = 59.0$ in. (1498.6 mm); $A_s = 8(0.723) = 5.784$ sq.in. (37.3184 cm²); $p = A_s/(bd) = 5.784/[32(59.0)] = 0.00306$.

Apply Eq. 61, or $f_{su} = 236,000(1 - 0.5 \times 0.00306 \times 236,000/5000) = 219,000$ lb/sq.in. (1,510,005.0 kPa). Also, $T_u = A_s f_{su} = 5.784(219,000) = 1,267,000$ lb (5,635,616.0 N). The concrete area under stress $= 1,267,000/[0.85(5,000)] = 298$ sq.in. (1922.7 cm²). This is the shaded area in Fig. 46, as the following calculation proves: $32(9.53) - 4.59(1.53) = 305 - 7 = 298$ sq.in. (1922.7 cm²).

Locate the centroidal axis of the stressed area, or $m = [305(4.77) - 7(9.53 - 0.51)]/298 = 4.67$ in. (118.618 mm); $M_u = \theta T_u jd = 0.90(1,267,000)(59.0 - 4.67) = 61,950,000$ in.·lb (6,999,111.0 N·m).

Calculate the steel index to ascertain that it is below the limit imposed by the ACI *Code*. Refer to Fig. 46. Or, area of $ABCD = 8(9.53) = 76.24$ sq.in. (491.900 cm²). The steel area A_{sr} that is required to balance the force on this web strip is $A_{sr} = 5.784(76.24)/298 = 1.48$ sq.in. (9.549 cm²); $q = A_{sr}f_{su}/(b'df_c') = 1.48(219,000)/[8(59.0)(5000)] = 0.137 < 0.30$. This is acceptable.

9. Calculate the required ultimate-moment capacity as given by the ACI Code
Thus, $w_u = 1.5(892 + 1160) + 1.8(1000) = 4878$ lb/lin ft (71,189.0 N/m); M_u required $= (⅛)(4878)(90)^2(12) = 59,270,000$ in.·lb (6,696,324.6 N·m). This is acceptable. The member is therefore adequate with respect to its ultimate-moment capacity.

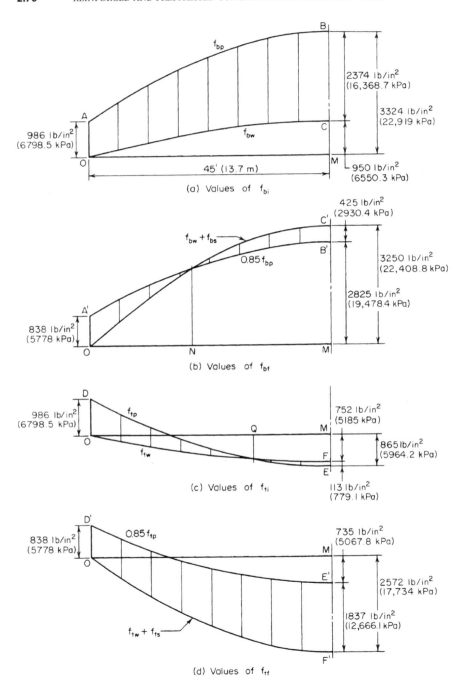

FIGURE 45

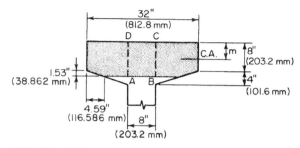

FIGURE 46. Concrete area under stress at ultimate load.

10. Design the web reinforcement
Follow the procedure given in step 8 of the previous calculation procedure.

11. Design the end block
This is usually done by applying isobar charts to evaluate the tensile stresses caused by the concentrated prestressing forces. Refer to Winter et al.—*Design of Concrete Structures*, McGraw-Hill.

12. Compute the camber at transfer
Referring to earlier procedures in this section, we see that $E_c I = 3.644(10)^6(394{,}800) = 1.44 \times 10^{12}$ lb·in² (4.132×10^9 N·m²). Also, $\Delta_w = -5(892)(90)^4(1728)/[384(1.44)(10)^{12}] = -0.91$ in. (-23.11 mm). Apply Eq. 60, or $\Delta_p = 5(844{,}000)(31.6)(90)^2(144)/[48(1.44)(10)^{12}] = 2.25$ in. (57.15 mm); $\Delta_i = 2.25 - 0.91 = 1.34$ in. (34.036 mm).

PROPERTIES OF A PARABOLIC ARC

Figure 47 shows the literal values of the coordinates at the ends and at the center of the parabolic arc $P_1 P_2 P_3$. Develop equations for y, dy/dx, and d^2y/dx^2 at an arbitrary point P. Find the slope of the arc at P_1 and P_3 and the coordinates of the summit S. (This information is required for the analysis of beams having parabolic trajectories.)

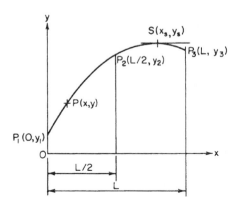

FIGURE 47. Parabolic arc.

Calculation Procedure:

1. Select a slope for the arc
Let m denote the slope of the arc.

2. Present the results
The equations are

$$y = 2(y_1 - 2y_2 + y_3)\left(\frac{x}{L}\right)^2 - (3y_1 - 4y_2 + y_3)\frac{x}{L} + y_1 \tag{66}$$

$$m = \frac{dy}{dx} = 4(y_1 - 2y_2 + y_3)\left(\frac{x}{L^2}\right) - \frac{3y_1 - 4y_2 + y_3}{L} \tag{67}$$

$$\frac{dm}{dx} = \frac{d^2y}{dx^2} = \frac{4}{L^2}(y_1 - 2y_2 + y_3) \tag{68}$$

$$m_1 = \frac{-(3y_1 - 4y_2 + y_3)}{L} \tag{69a}$$

$$m_3 = \frac{y_1 - 4y_2 + 3y_3}{L} \tag{69b}$$

$$x_s = \frac{(L/4)(3y_1 - 4y_2 + y_3)}{y_1 - 2y_2 + y_3} \tag{70a}$$

$$y_s = \frac{-(1/8)(3y_1 - 4y_2 + y_3)^2}{y_1 - 2y_2 + y_3} + y_1 \tag{70b}$$

ALTERNATIVE METHODS OF ANALYZING A BEAM WITH PARABOLIC TRAJECTORY

The beam in Fig. 48 is subjected to an initial prestressing force of 860 kips (3825.3 kN) on a parabolic trajectory. The eccentricities at the left end, midspan, and right end, respectively, are $e_a = 1$ in. (25.4 mm); $e_m = 30$ in. (762.0 mm); $e_b = -3$ in. (−76.2 mm). Evaluate the prestress shear and prestress moment at section C (a) by applying the properties of the trajectory at C; (b) by considering the prestressing action of the steel on the concrete in the interval AC.

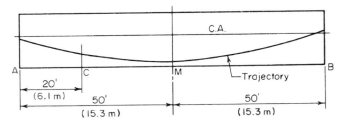

FIGURE 48

Calculation Procedure:

1. Compute the eccentricity and slope of the trajectory at C
Use Eqs. 66 and 67. Let m denote the slope of the trajectory. This is positive if the trajectory slopes downward to the right. Thus $e_a - 2e_m + e_b = 1 - 60 - 3 = -62$ in. (1574.8 mm); $3e_a - 4e_m + e_b$, $= 3 - 120 - 3 = -120$ in. (−3048 mm); $e_m = 2(-62)(20/100)^2 + 120(20/100) + 1 = 20.04$ in. (509.016 mm); $m_c = 4(-62/12)(20/100^2) - (-120/12 \times 100) = 0.0587$.

2. Compute the prestress shear and moment at C
Thus $V_{pc} = -m_c F_i = -0.0587(860,000) = -50,480$ lb (−224,535.0 N); $M_{pc} = -F_i e = -860,000(20.04) = -17,230,000$ in.·lb (−1,946,645.4 N·m). This concludes the solution to part a.

3. Evaluate the vertical component w of the radial force on the concrete in a unit longitudinal distance
An alternative approach to this problem is to analyze the forces that the tendons exert on the concrete in the interval AC, namely, the prestressing force transmitted at the end and the radial forces resulting from curvature of the tendons.

Consider the component w to be positive if directed downward. In Fig. 49, $V_{pr} - V_{pq} = -F_i(m_r - m_q)$; therefore, $\Delta V_p / \Delta x = -F_i \Delta m / \Delta x$. Apply Eq. 68: $dV_p/dx = -F_i dm/dx = -(4F_i/L^2)(e_a - 2e_m + e_b)$; but $dV_p/dx = -w$. Therefore,

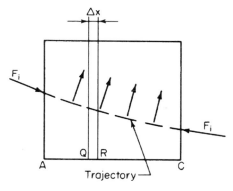

FIGURE 49. Free-body diagram of concrete.

$$w = F_i \frac{dm}{dx} = \left(\frac{4F_i}{L^2}\right)(e_a - 2e_m + e_b) \qquad (71)$$

This result discloses that when the trajectory is parabolic, w is uniform across the span. The radial forces are always directed toward the center of curvature, since the tensile forces applied at their ends tend to straighten the tendons. In the present instance, $w = (4F_i/100^2)(-62/12) = -0.002067F_i$ lb/lin ft (−0.00678F_i N/m).

4. Find the prestress shear at C
By Eq. 69a, $m_a = -[-120/(100 \times 12)] = 0.1$; $V_{pa} = -0.1F_i$; $V_{pc} = V_{pa} - 20w = F_i(-0.1 + 20 \times 0.002067) = -0.0587F_i = -50,480$ lb (−224,535.0 N).

5. Find the prestress moment at C
Thus, $M_{pc} = M_{pa} + V_{pa}(240) - 20w(120) = F_i - 0.1 \times 240 + (20 \times 0.002067 \times 120) = -20.04F_i = -17,230,000$ in.·lb (1,946,645.4 N·m).

PRESTRESS MOMENTS IN A CONTINUOUS BEAM

The continuous prismatic beam in Fig. 50 has a prestressing force of 96 kips (427.0 kN) on a parabolic trajectory. The eccentricities are $e_a = -0.40$ in. (−10.16 mm); $e_d = +0.60$ in. (15.24 mm); $e_b = -1.20$ in. (−30.48 mm); $e_e = +0.64$ in. (16.256 mm); $e_c = -0.60$ in. (−15.24 mm). Construct the prestress-moment diagram for this member, indicating all significant values.

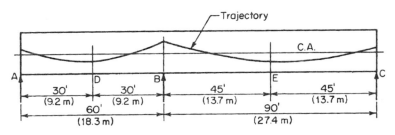

FIGURE 50

Calculation Procedure:

1. Find the value of $wL^2/4$ for each span by applying Eq. 71
Refer to Fig. 51. Since members AB and BC are constrained to undergo an identical rotation at B, there exists at this section a bending moment M_{kb} in addition to that resulting from the eccentricity of F_i. The moment M_{kb} induces reactions at the supports. Thus, at every section of the beam there is a moment caused by continuity of the member as well as the moment $-F_i e$. The moment M_{kb} is termed the continuity moment; its numerical value is directly proportional to the distance from the given section to the end support. The continuity moment may be evaluated by adopting the second method of solution in the previous calculation procedure, since this renders the continuous member amenable to analysis by the theorem of three moments or moment distribution.

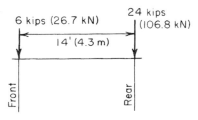

FIGURE 51. Free-body diagram of concrete.

Determine $wL^2/4$ for each span: span AB, $w_1 L_1^2/4 = F_i(-0.40 - 1.20 - 1.20) = -2.80F_i$ in.·lb $(-0.3163F_i$ N·m); span BC, $w_2 L_2^2/4 = F_i(-1.20 - 1.28 - 0.60) = -3.08F_i$ in.·lb $(-0.3479F_i$ N·m).

2. Determine the true prestress moment at B in terms of F_1
Apply the theorem of three moments; by subtraction, find M_{kb}. Thus, $M_{pa}L_1 + 2M_{pb}(L_1 + L_2) + M_{pc}L_2 = -w_1 L_1^3/4 - w_2 L_2^3/4$. Substitute the value of L_1 and L_2, in feet (meters), and divide each term by F_i, or $0.40(60) + (2M_{pb} \times 150)/F_i + 0.60(90) = 2.80(60) + 3.08(90)$. Solving gives $M_{pb} = 1.224F_i$ in.·lb $(0.1383F_i$ N·m). Also, $M_{kb} = M_{pb} - (-F_i e_b) = F_i(1.224 - 1.20) = 0.024F_i$. Thus, the continuity moment at B is positive.

3. Evaluate the prestress moment at the supports and at midspan
Using foot-pounds (Newton-meters) in the moment evaluation yields $M_{pa} = 0.40(96,000)/12 = 3200$ ft·lb $(4339.2$ N·m); $M_{pb} = 1.224(96,000)/12 = 9792$ ft·lb $(13,278$ N·m); $M_{pc} = 0.60(96,000)/12 = 4800$ ft·lb $(6508.0$ N·m); $M_{pd} = -F_j e_d + M_{kd} = F_i(-0.60 + \frac{1}{2} \times 0.024)/12 = -4704$ ft·lb $(-6378$ N·m); $M_{pe} = F_i(-0.64 + \frac{1}{2} \times 0.024)/12 = -5024$ ft·lb $(-6812$ N·m).

4. Construct the prestress-moment diagram
Figure 52 shows this diagram. Apply Eq. 70 to locate and evaluate the maximum negative moments. Thus, $AF = 25.6$ ft $(7.80$ m); $BG = 49.6$ ft $(15.12$ m); $M_{pf} = -4947$ ft·lb $(-6708$ N·m); $M_{pg} = -5151$ ft·lb $(-6985$ N·m).

PRINCIPLE OF LINEAR TRANSFORMATION

For the beam in Fig. 50, consider that the parabolic trajectory of the prestressing force is displaced thus: e_a and e_c are held constant as e_b is changed to -2.0 in. (-50.80 mm), the eccentricity at any intermediate section being decreased algebraically by an amount directly proportional to the distance from that section to A or C. Construct the prestress-moment diagram.

Calculation Procedure:

1. Compute the revised eccentricities

The modification described is termed a *linear transformation* of the trajectory. Two methods are presented. Steps 1 through 4 comprise method 1; the remaining steps comprise method 2.

The revised eccentricities are $e_a = -0.40$ in. (-10.16 mm); $e_d = +0.20$ in. (5.08 mm); $e_b = -2.00$ in. (-50.8 mm); $e_e = +0.24$ in. (6.096 mm); $e_c = -0.60$ in. (-15.24 mm).

2. Find the value of $wL^2/4$ for each span

Apply Eq. 71: span AB, $w_1L_1^2/4 = F_i(-0.40 - 0.40 - 2.00) = -2.80F_i$; span BC, $w_2L_2^2/4 = F_i(-2.00 - 0.48 - 0.60) = -3.08F_i$.

These results are identical with those obtained in the previous calculation procedure. The change in e_b is balanced by an equal change in $2e_d$ and $2e_e$.

3. Determine the true prestress moment at B by applying the theorem of three moments; then find M_{kb}

Refer to step 2 in the previous calculation procedure. Since the linear transformation of the trajectory has not affected the value of w_1 and w_2, the value of M_{pb} remains constant. Thus, $M_{kb} = M_{pb} - (-F_ie_b) = F_i(1.224 - 2.0) = -0.776F_i$.

4. Evaluate the prestress moment at midspan

Thus, $M_{pd} = -F_ie_d + M_{kd} = F_i(-0.20 - \frac{1}{2} \times 0.776)/12 = -4704$ ft·lb (-6378.6 N·m); $M_{pe} = F_i(-0.24 - \frac{1}{2} \times 0.776)/12 = -5024$ ft·lb (-6812.5 N·m).

These results are identical with those in the previous calculation procedure. The change in the eccentricity moment is balanced by an accompanying change in the continuity moment. Since three points determine a parabolic arc, the prestress moment diagram coincides with that in Fig. 52. This constitutes the solution by method 1.

5. Evaluate the prestress moments

Do this by replacing the prestressing system with two hypothetical systems that jointly induce eccentricity moments identical with those of the true system.

Let e denote the original eccentricity of the prestressing force at a given section and Δe the change in eccentricity that results from the linear transformation. The final eccentricity moment is $-F_i(e + \Delta e) = -(F_ie + F_i\Delta e)$.

Consider the beam as subjected to two prestressing forces of 96 kips (427.0 kN) each. One has the parabolic trajectory described in the previous calculation procedure; the other has the linear trajectory shown in Fig. 53, where $e_a = 0$, $e_b = -0.80$ in. (-20.32 mm), and $e_c = 0$. Under the latter prestressing system, the tendons exert three forces on the concrete—one at each end and one at the deflection point above the interior support caused by the change in direction of the prestressing force.

The horizontal component of the prestressing force is considered equal to the force itself; it therefore follows that the force acting at the deflection point has no horizontal component.

Since the three forces that the tendons exert on the concrete are applied directly at the supports, their vertical components do not induce bending. Similarly, since the

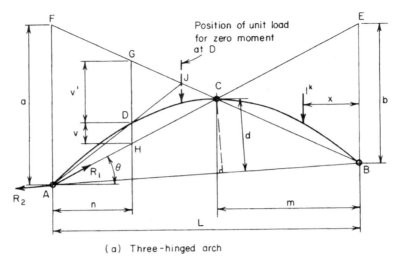

(a) Three-hinged arch

FIGURE 52. Prestress moment diagram.

forces at A and C are applied at the centroidal axis, their horizontal components do not induce bending. Consequently, the prestressing system having the trajectory shown in Fig. 53 does not cause any prestress moments whatsoever. The prestress moments for the beam in the present instance are therefore identical with those for the beam in the previous calculation procedure.

The second method of analysis is preferable to the first because it is general. The first method demonstrates the equality of prestress moments before and after the linear transformation where the trajectory is parabolic; the second method demonstrates this equality without regard to the form of trajectory.

In this calculation procedure, the extremely important principle of linear transformation for a two-span continuous beam was developed. This principle states: The prestress moments remain constant when the trajectory of the prestressing force is transformed linearly. The principle is frequently applied in plotting a trial trajectory for a continuous beam.

Two points warrant emphasis. First, in a linear transformation, the eccentricities at the end supports remain constant. Second, the hypothetical prestressing systems introduced in step 5 are equivalent to the true system solely with respect to bending stresses; the axial stress F_i/A under the hypothetical systems is double that under the true system.

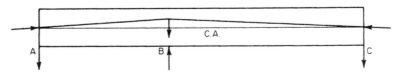

FIGURE 53. Hypothetical prestressing system and forces exerted on concrete.

EFFECT OF VARYING ECCENTRICITY AT END SUPPORT

For the beam in Fig. 50, consider that the parabolic trajectory in span AB is displaced thus: e_b is held constant as e_a is changed to -0.72 in. (-18.288 mm), the eccentricity at every intermediate section being decreased algebraically by an amount directly proportional to the distance from that section to B. Compute the prestress moment at the supports and at midspan caused by a prestressing force of 96 kips (427.0 kN).

Calculation Procedure:

1. Apply the revised value of e_a; repeat the calculations of the earlier procedure

Thus, $M_{pa} = 5760$ ft·lb (7810.6 N·m); $M_{pd} = -3680$ ft·lb (-4990.1 N·m); $M_{pb} = 9280$ ft·lb (12,583.7 N·m); $M_{pe} = -5280$ ft·lb (-7159.7 N·m); $M_{pc} = 4800$ ft·lb (6508.8 N·m).

The change in prestress moment caused by the displacement of the trajectory varies linearly across each span. Figure 54 compares the original and revised moments along AB. This constitutes method 1.

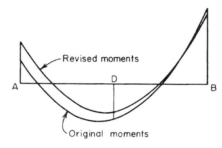

FIGURE 54. Prestress-moment diagrams.

2. Replace the prestressing system with two hypothetical systems that jointly induce eccentricity moments identical with those of the true system

This constitutes method 2. For this purpose, consider the beam to be subjected to two prestressing forces of 96 kips (427.0 kN) each. One has the parabolic trajectory described in the earlier procedure; the other has a trajectory that is linear in each span, the eccentricities being $e_a = -0.72 - (-0.40) = -0.32$ in. (-8.128 mm), $e_b = 0$, and $e_c = 0$.

3. Evaluate the prestress moments induced by the hypothetical system having the linear trajectory

The tendons exert a force on the concrete at A, B, and C, but only the force at A causes bending moment.

Thus, $M_{pa} = -F_i e_a = -96,000(-0.32)/12 = 2560$ ft·lb (3471.4 N·m). Also, $M_{pa}L_1 + 2M_{pb}(L_1 + L_2) + M_{pc}L_2 = 0$. But $M_{pc} = 0$; therefore, $M_{pb} = -512$ ft·lb (-694.3 N·m); $M_{pd} = \frac{1}{2}(2560 - 512) = 1024$ ft·lb (1388.5 N·m); $M_{pe} = \frac{1}{2}(-512) = -256$ ft·lb (-347.1 N·m).

4. Find the true prestress moments by superposing the two hypothetical systems

Thus $M_{pa} = 3200 + 2560 = 5760$ ft·lb (7810.6 N·m); $M_{pd} = 4704 + 1024 = 3680$ ft·lb (-4990.1 N·m); $M_{pb} = 9792 - 512 = 9280$ ft·lb (12,583.7 N·m); $M_{pe} = 5024 - 256 = -5280$ ft·lb (-7159.7 N·m); $M_{pc} = 4800$ ft·lb (6508.8 N·m).

DESIGN OF TRAJECTORY FOR A TWO-SPAN CONTINUOUS BEAM

A T beam that is continuous across two spans of 120 ft (36.6 m) each is to carry a uniformly distributed live load of 880 lb/lin ft (12,842.6 N/m). The cross section has these properties: $A = 1440$ sq.in. (9290.8 cm^2); $I = 752,000$ in^4 (3130.05 dm^4); $y_b = 50.6$ in. (1285.24 mm);

$y_t = 23.4$ in. (594.36 mm). The allowable stresses are: *initial*, +2400 and −60 lb/sq.in. (+16,548.0 and −413.7 kPa); *final*, +2250 and −60 lb/sq.in. (+15,513.8 and −413.7 kPa). Assume that the minimum possible distance from the extremity of the section to the centroidal axis of the prestressing steel is 9 in. (228.6 mm). Determine the magnitude of the prestressing force, and design the parabolic trajectory (*a*) using solely prestressed reinforcement; (*b*) using a combination of prestressed and non-prestressed reinforcement.

Calculation Procedure:

1. Compute the section moduli, kern distances, and beam weight

For part *a*, an exact design method consists of these steps: First, write equations for the prestress moment, beam-weight moment, maximum and minimum potential superimposed-load moment, expressing each moment in terms of the distance from a given section to the adjacent exterior support. Second, apply these equations to identify the sections at which the initial and final stresses are critical. Third, design the prestressing system to restrict the critical stresses to their allowable range. Whereas the exact method is not laborious when applied to a prismatic beam carrying uniform loads, this procedure adopts the conventional, simplified method for illustrative purposes. This consists of dividing each span into a suitable number of intervals and analyzing the stresses at each boundary section.

For simplicity, set the eccentricity at the ends equal to zero. The trajectory will be symmetric about the interior support, and the vertical component *w* of the force exerted by the tendons on the concrete in a unit longitudinal distance will be uniform across the entire length of member. Therefore, the prestress-moment diagram has the same form as the bending-moment diagram of a nonprestressed prismatic beam continuous over two equal spans and subjected to a uniform load across its entire length. It follows as a corollary that the prestress moments at the boundary sections previously referred to have specific *relative values* although their absolute values are functions of the prestressing force and its trajectory.

The following steps constitute a methodical procedure: Evaluate the relative prestress moments, and select a trajectory having ordinates directly proportional to these moments. The trajectory thus fashioned is concordant. Compute the prestressing force required to restrict the stresses to the allowable range. Then transform the concordant trajectory linearly to secure one that lies entirely within the confines of the section. Although the number of satisfactory concordant trajectories is infinite, the one to be selected is that which requires the minimum prestressing force. Therefore, the selection of the trajectory and the calculation of F_i are blended into one operation.

Divide the left span into five intervals, as shown in Fig. 55. (The greater the number of intervals chosen, the more reliable are the results.)

Computing the moduli, kern distances, and beam weight gives $S_b = 14,860$ in^3 (243,555.4 cm^3); $S_t = 32,140$ in^3 (526,774.6 cm^3); $k_b = 22.32$ in. (566.928 mm); $k_t = 10.32$ in. (262,128 mm); $w_w = 1500$ lb/lin ft (21,890.9 N/m).

2. Record the bending-moment coefficients C_1, C_2, and C_3

Use Table 4 to record these coefficients at the boundary sections. The subscripts refer to these conditions of loading: 1, load on entire left span and none on right span; 2, load on entire right span and none on left span; 3, load on entire length of beam.

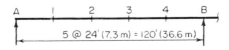

FIGURE 55. Division of span into intervals.

TABLE 4. Calculations for Two-Span Beam: Part a

Section	1	2	3	4	B
1 C_1	+0.0675	+0.0950	+0.0825	+0.0300	−0.0625
2 C_2	−0.0125	−0.0250	−0.0375	−0.0500	−0.0625
3 C_3	+0.0550	+0.0700	+0.0450	−0.0200	−0.1250
4 f_{bw}, lb/sq.in. (kPa)	−959 (−6,611)	−1,221 (−8,418)	−785 (−5,412)	+349 (+2406)	+2,180 (+15,029)
5 f_{bs1}, lb/sq.in. (kPa)	−691 (−4,764)	−972 (−6,701)	−844 (−5,819)	−307 (−2,116)	+640 (+4,412)
6 f_{bs2}, lb/sq.in. (kPa)	+128 (+882)	+256 (+1,765)	+384 (+2,647)	+512 (+3,530)	+640 (+4,412)
7 f_{tw}, lb/sq.in. (kPa)	+444 (+3,060)	+565 (+3895)	+363 (+2,503)	−161 (−1,110)	−1,008 (−6,949)
8 f_{ts1}, lb/sq.in. (kPa)	+319 (+2,199)	+450 (+3,102)	+390 (+2689)	+142 (+979)	−296 (−2,041)
9 f_{ts2}, lb/sq.in. (kPa)	−59 (−407)	−118 (−813)	−177 (−1,220)	−237 (−1,634)	−296 (−2,041)
10 e_{con}, in. (mm)	+17.19 (+436.6)	+21.87 (+555.5)	+14.06 (+357.1)	−6.25 (−158.8)	−39.05 (−991.9)
11 f_{bp}, lb/sq.in. (kPa)	+2,148 (+14,808)	+2,513 (+17,325)	+1,903 (+13,119)	+318 (+2,192)	−2,243 (−15,463)
12 f_{tp}, lb/sq.in. (kPa)	+185 (+128)	+16 (+110)	+298 (+2,054)	+1,031 (+7,108)	+2,215 (+15,270)
13 $0.85 f_{bp}$ lb/sq.in. (kPa)	+1,826 (+12,588)	+2,136 (+14,726)	+1,618 (+11,154)	+270 (+1,861)	−1,906 (−13,140)
14 $0.85 f_{tp}$ lb/sq.in. (kPa)	+157 (+1,082)	+14 (+97)	+253 (+1,744)	+876 (+6,039)	+1,883 (+12,981)

At midspan: $C_3 = +0.0625$ and $e_{con} = +19.53$ in. (496.1 mm)

To obtain these coefficients, refer to the AISC *Manual*, case 29, which represents condition 1. Thus, $R_1 = (7/16)wL$; $R_3 = -(1/16)wL$. At section 3, for example, $M_1 = (7/16)wL(0.6L) - \frac{1}{2}w(0.6L)^2 = [7(0.6) - 8(0.36)]wL^2/16 = 0.0825wL^2$; $C_1 = M_1/(wL^2) = +0.0825$.

To obtain condition 2, interchange R_1 and R_3. At section 3, $M_2 = -(1/16)wL(0.6L) = -0.0375wL^2$; $C_2 = -0.0375$; $C_3 = C_1 + C_2 = +0.0825 - 0.0375 = +0.0450$.

These moment coefficients may be applied without appreciable error to find the maximum and minimum potential live-load bending moments at the respective sections. The values of C_3 also represent the relative eccentricities of a concordant trajectory.

Since the gravity loads induce the maximum positive moment at section 2 and the maximum negative moment at section B, the prestressing force and its trajectory will be designed to satisfy the stress requirements at these two sections. (However, the stresses at all boundary sections will be checked.) The Magnel diagram for section 2 is similar to that in Fig. 37, but that for section B is much different.

3. Compute the value of C_3 at midspan
Thus, $C_3 = +0.0625$.

4. Apply the moment coefficients to find the gravity-load stresses
Record the results in Table 4. Thus $M_w = C_3(1500)(120)^2(12) = 259,200,000 C_3$ in.·lb ($29.3 C_3$ kN·m); $f_{bw} = -259,200,000 C_3/14,860 = -17,440 C_3$; $f_{bs1} = -10,230 C_1$; $f_{bs2} = -10,230 C_2$; $f_{tw} = 8065 C_3$; $f_{ts1} = 4731 C_1$; $f_{ts2} = 4731 C_2$.

2.80 REINFORCED AND PRESTRESSED CONCRETE ENGINEERING AND DESIGN

Since S_t far exceeds S_b, it is manifest that the prestressing force must be designed to confine the bottom-fiber stresses to the allowable range.

5. Consider that a concordant trajectory has been plotted; express the eccentricity at section B relative to that at section 2
Thus, $e_b/e_2 = -0.1250/+0.0700 = -1.786$; therefore, $e_b = -1.786e_2$.

6. Determine the allowable range of values of f_{bp} at sections 2 and B
Refer to Fig. 56. At section 2, $f_{bp} \leq +3621$ lb/sq.in. (+24,966.8 kPa), Eq. a; $0.85 f_{bp} \geq 1221 + 972 - 60$; therefore, $f_{bp} \geq +2509$ lb/sq.in. (+17,299.5 kPa), Eq b. At section B, $f_{bp} \geq -2240$ lb/sq.in. (-15,444.8 kPa), Eq. c; $0.85 f_{bp} \leq -(2180 + 1280) + 2250$; $f_{bp} \leq -1424$ lb/sq.in. (-9818.5 kPa), Eq. d.

7. Substitute numerical values in Eq. 56, expressing e_b in terms of e_2
The values obtained are $1/F_i \leq (k_t + e_2)/(3621 S_b)$, Eq a'; $1/F_i \leq (k_t + e_2)/(2509 S_b)$, Eq. b'; $1/F_t \geq (1.786 e_2 - k_t)/(2240 S_b)$, Eq. c'; $1/F_i \leq (1.786 e_2 - k_t)/(1424 S_b)$, Eq. d'.

8. Obtain the composite Magnel diagram
Considering the relations in step 7 as equalities, plot the straight lines representing them to obtain the composite Magnel diagram in Fig. 57. The slopes of the lines have these relative values: $m_a = 1/3621$; $m_b = 1/2509$; $m_c = 1.786/2240 = 1/1254$; $m_d = 1.786/1424 = 1/797$. The shaded area bounded by these lines represents the region of permissible sets of values of e_2 and $1/F_i$.

9. Calculate the minimum allowable value of F_i and the corresponding value of e_2
In the composite Magnel diagram, this set of values is represented by point A. Therefore, consider Eqs. b' and c' as equalities, and solve for the unknowns. Or, $(10.32 + e_2)/2509 = (1.786 e_2 - 10.32)/2240$; solving gives $e_2 = 21.87$ in. (555.5 mm) and $F_i = 1,160,000$ lb (5,159,680.0 N).

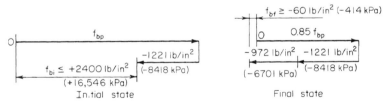

(a) Limiting values of f_{bp} at section 2

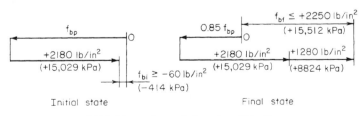

(b) Limiting values of f_{bp} at section B

FIGURE 56

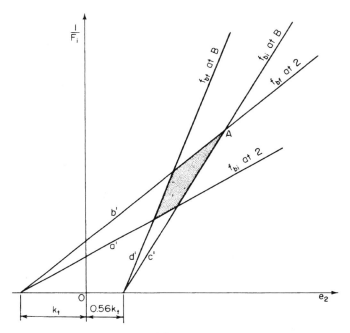

FIGURE 57. Composite Magnel diagram.

10. Plot the concordant trajectory
Do this by applying the values of C_3 appearing in Table 4; for example, $e_1 = +21.87(0.0550)/0.0700 = +17.19$ in. (436.626 mm). At midspan, $e_m = +21.87(0.0625)/0.0700 = +19.53$ in. (496.062 mm).

Record the eccentricities on line 10 of the table. It is apparent that this concordant trajectory is satisfactory in the respect that it may be linearly transformed to one falling within the confines of the section; this is proved in step 14.

11. Apply Eq. 56 to find f_{bp} and f_{tp}
Record the results in Table 4. For example, at section 1, $f_{bp} = 1,160,000(10.32 + 17.19)/14,860 = +2148$ lb/sq.in. (+14,810.5 kPa); $f_{tp} = 1,160,000(22.32 - 17.19)/32,140 = +185$ lb/sq.in. (+1275.6 kPa).

12. Multiply the values of f_{bp} and f_{tp} by 0.85, and record the results
These results appear in Table 4.

13. Investigate the stresses at every boundary section
In calculating the final stresses, apply the live-load stress that produces a more critical condition. Thus, at section 1, $f_{bi} = -959 + 2148 = +1189$ lb/sq.in. (+8198.2 kPa); $f_{bf} = -959 - 691 + 1826 = +176$ lb/sq.in. (+1213.6 kPa); $f_{ti} = +444 + 185 = +629$ lb/sq.in. (+4337.0 kPa); $f_{tf} = +444 + 319 + 157 = +920$ lb/sq.in. (+6343.4 kPa). At section 2: $f_{bi} = -1221 + 2513 = +1292$ lb/sq.in. (+8908.3 kPa); $f_{bf} = -1221 - 972 + 2136 = -57$ lb/sq.in. (-393.0 kPa); $f_{ti} = +565 + 16 = +581$ lb/sq.in. (+4006.0 kPa); $f_{tf} = +565 + 450 + 14 = +1029$ lb/sq.in. (+7095.0 kPa). At section 3: $f_{bi} = -785 + 1903 = +1118$ lb/sq.in. (+7706.8 kPa); $f_{bf} = -785 - 844 + 1618 = -11$ lb/sq.in. (-75.8 kPa); $f_{ti} = +363 + 298 = +661$ lb/sq.in. (+4558.0 kPa); $= +363 + 390 + 253 = +1006$ lb/sq.in.

(+6936.4 kPa). At section 4: $f_{bi} = +349 + 318 = +667$ lb/sq.in. (+4599.0 kPa); $f_{bf} = +349 - 307 + 270 = +312$ lb/sq.in. (+2151.2 kPa); or $f_{bf} = +349 + 512 + 270 = +1131$ lb/sq.in. (7798.2 kPa); $f_{ti} = -161 + 1031 = +870$ lb/sq.in. (+5998.7 kPa); $f_{tf} = -161 - 237 + 876 = +478$ lb/sq.in. (+3295.8 kPa), or $f_{tf} = -161 + 142 + 876 = +857$ lb/sq.in. (+5909.0 kPa). At section B: $f_{bi} = +2180 - 2243 = -63$ lb/sq.in. (-434.4 kPa); $f_{bf} = +2180 + 1280 - 1906 = +1554$ lb/sq.in. (+10,714.8 kPa); $f_{ti} = -1008 + 2215 = +1207$ lb/sq.in. (+8322.3 kPa); $f_{tf} = -1008 - 592 + 1883 = +283$ lb/sq.in. (+1951.3 kPa).

In all instances, the stresses lie within the allowable range.

14. Establish the true trajectory by means of a linear transformation

The imposed limits are $e_{max} = y_b - 9 = 41.6$ in. (1056.6 mm), $e_{min} = -(y_t - 9) = -14.4$ in. (-365.76 mm).

Any trajectory that falls between these limits and that is obtained by linearly transforming the concordant trajectory is satisfactory. Set $e_b = -14$ in. (-355.6 mm), and compute the eccentricity at midspan and the maximum eccentricity.

Thus, $e_m = +19.53 + \frac{1}{2}(39.05 - 14) = +32.06$ in. (814.324 mm). By Eq. 70b, $e_s = -(\frac{1}{8})(-4 \times 32.06 - 14)^2/(-2 \times 32.06 - 14) = +32.4$ in. (+823.0 mm) < 41.6 in. (1056.6 mm). This is acceptable. This constitutes the solution to part a of the procedure. Steps 15 through 20 constitute the solution to part b.

15. Assign eccentricities to the true trajectory, and check the maximum eccentricity

The preceding calculation shows that the maximum eccentricity is considerably below the upper limit set by the beam dimensions. Refer to Fig. 57. If the restrictions imposed by line c' are removed, e_2 may be increased to the value corresponding to a maximum eccentricity of 41.6 in. (1056.6 mm), and the value of F_i is thereby reduced. This revised set of values will cause an excessive initial tensile stress at B, but the condition can be remedied by supplying nonprestressed reinforcement over the interior support. Since the excess tension induced by F_i extends across a comparatively short distance, the savings accruing from the reduction in prestressing force will more than offset the cost of the added reinforcement.

Assigning the following eccentricities to the true trajectory and checking the maximum eccentricity by applying Eq. 70b, we get $e_a = 0$; $e_m = +41$ in. (1041.4 mm); $e_b = -14$ in. (-355.6 mm); $e_s = -(\frac{1}{8})(-4 \times 41 - 14)^2/(-2 \times 41 - 14) = +41.3$ in. (1049.02 mm). This is acceptable.

16. To analyze the stresses, obtain a hypothetical concordant trajectory by linearly transforming the true trajectory

Let y denote the upward displacement at B. Apply the coefficients C_3 to find the eccentricities of the hypothetical trajectory. Thus, $e_m/e_b = (41 - \frac{1}{2}y)/(-14 - y) = +0.0625/-0.1250$; $y = 34$ in. (863.6 mm); $e_a = 0$; $e_m = +24$ in. (609.6 mm); $e_b = -48$ in. (-1219.2 mm); $e_1 = -48(+0.0550)/-0.1250 = +21.12$ in. (536.448 mm); $e_2 = +26.88$ in. (682.752 mm); $e_3 = +17.28$ in. (438.912 mm); $e_4 = -7.68$ in. (-195.072 mm).

17. Evaluate F_i by substituting in relation (b') of step 7

Thus, $F_i = 2509(14,860)/(10.32 + 26.88) = 1,000,000$ lb (4448 kN). Hence, the introduction of nonprestressed reinforcement served to reduce the prestressing force by 14 percent.

18. Calculate the prestresses at every boundary section; then find the stresses at transfer and under design load

Record the results in Table 5. (At sections 1 through 4, the final stresses were determined by applying the values on lines 5 and 8 in Table 4. The slight discrepancy between the final stress at 2 and the allowable value of -60 lb/sq.in. (-413.7 kPa) arises from the degree of precision in the calculations.)

TABLE 5. Calculations for Two-Span Continuous Beam: Part *b*

Section	1	2	3	4	B
e_{con}, in. (mm)	+21.12	+26.88	+17.28	−7.68	−48.00
	(536.4)	(+682.8)	(+438.9)	(−195.1)	(−1,219.2)
f_{bp}, lb/sq.in. (kPa)	+2,116	+2,503	+1,857	+178	−2,535
	(+14,588)	(+17,256)	(+12,802)	(+1,227)	(−17,476)
f_{tp}, lb/sq.in. (kPa)	+37	−142	+157	+933	+2188
	(+255)	(−979)	(+1,082)	(+6,660)	(+15,084)
$0.85 f_{bp}$, lb/sq.in. (kPa)	+1,799	+2,128	+1,578	+151	−2,155
	(+12,402)	(+14,670)	(+10,879)	(+1,041)	(−14,857)
$0.85 f_{tp}$, lb/sq.in. (kPa)	+31	−121	+133	+793	+1,860
	(+214)	(−834)	(+917)	(+5,467)	(+12,823)
f_{bi}, lb/sq.in. (kPa)	+1,157	+1,282	+1,072	+527	−355
	(+7,976)	(+8,838)	(+7,390)	(+3,633)	(−2,447)
f_{bf}, lb/sq.in. (kPa)	+149	−65	−51	+193	+1,305
	(+1,027)	(−448)	(−352)	(+1,331)	(+8,997)
f_{ti}, lb/sq.in. (kPa)	+481	+423	+520	+772	+1,180
	(+3,316)	(+2,916)	(+3,585)	(+5,322)	(+8,135)
f_{tf}, lb/sq.in. (kPa)	+794	+894	+886	+774	+260
	(+5,474)	(+6,163)	(+6108)	(+5,336)	(+1,792)

With the exception of f_{bi} at B, all stresses at the boundary sections lie within the allowable range.

19. Locate the section at which f_{bi} = −60 lb/sq.in. (−413.7 kPa)

Since f_{bp} and f_{bw} vary parabolically across the span, their sum f_{bi} also varies in this manner. Let x denote the distance from the interior support to a given section. Apply Eq. 66 to find the equation for f_{bi} using the initial-stress values at sections B, 3, and 1. Or, $−355 − 2 × 1072 + 1157 = −1342$ (−9253.1 kPa); $3(−355) − 4(1072) + 1157 = −4196$(−28,931.4 kPa); $f_{bi} = −2684(x/96)^2 + 4196 x/96 − 355$. When $f_{bi} = −60(−413.7)$, $x = 7.08$ ft (2.15 m). The tensile stress at transfer is therefore excessive in an interval of only 14.16 ft (4.32 m).

20. Design the nonprestressed reinforcement over the interior support

As in the preceding procedures, the member must be investigated for ultimate-strength capacity. The calculation pertaining to any quantity that varies parabolically across the span may be readily checked by verifying that the values at uniformly spaced sections have equal "second differences." For example, with respect to the values of f_{bi} recorded in Table 5, the verification is:

$$+1157 \quad +1282 \quad +1072 \quad +527 \quad -355$$

$$-125 \quad +210 \quad +545 \quad +882$$

$$+335 \quad +335 \quad +337$$

The values on the second and third lines represent the differences between successive values on the preceding line.

STEEL BEAM ENCASED IN CONCRETE

A concrete floor slab is to be supported by steel beams spaced 10 ft (3.05 m) on centers and having a span of 28 ft 6 in. (8.69 m). The beams will be encased in concrete with a minimum cover of 2 in. (50.8 mm) all around; they will remain unshored during construction. The slab has been designed as 4½ in. (114.3 mm) thick, with $f_c' = 3000$ lb/sq.in. (20.7 MPa). The loading includes the following: live load, 120 lb/sq.ft.(5.75 kPa); finished floor and ceiling, 25 lb/sq.ft. (1.2 kPa). The steel beams have been tentatively designed as W16 × 40. Review the design.

Calculation Procedure:

1. Record the relevant properties of the section and the allowable flexural stresses

In accordance with the AISC *Specification*, the member may be designed as a composite steel-and-concrete beam, reliance being placed on the natural bond of the two materials to obtain composite action. Refer to Sec. 1 for the design of a composite bridge member. In the design of a composite building member, the effects of plastic flow are usually disregarded. Since the slab is poured monolithically, the composite member is considered continuous. Apply the following equations in computing bending moments in the composite beams: at midspan, $M = (1/20)wL^2$; at support, $M = (1/12)wL^2$.

The subscripts c, ts, and bs refer to the extreme fiber of concrete, top of steel, and bottom of steel, respectively. The superscripts c and n refer to the composite and noncomposite sections, respectively.

Record the properties of the W16 × 40: $A = 11.77$ sq.in. (75.94 cm^2); $d = 16.00$ in. (406.4 mm); $I = 515.5$ in^4 (21.457 cm^4); $S = 64.4$ in^3 (1055.3 cm^3); flange width = 7 in. (177.8 mm). By the AISC *Specification*, $f_s = 24,000$ lb/sq.in. (165.5 MPa). By the ACI *Code*, $n = 9$ and $f_c = 1350$ lb/sq.in. (9306.9 kPa).

2. Transform the composite section in the region of positive moment to an equivalent section of steel; compute the section moduli

Refer to Fig. 58a and the AISC *Specification*. Use the gross concrete area. Then the effective flange width = ¼L = ¼(28.5)12 = 85.5 in. (2172 mm); spacing of beams = 120 in. (3048 mm); $16t + 11 = 16(4.5) + 11 = 83$ in. (2108 mm); this governs. Transformed width = 83/9 = 9.22 in. (234.2 mm).

Assume that the neutral axis lies within the flange, and take static moments with respect to this axis; or ½(9.22y^2) − 11.77(10 − y) = 0; $y = 3.93$ in. (99.8 mm).

Compute the moment of inertia. Slab: (1/3)9.22(3.93)^3 = 187 in^4 (7783.5 cm^4). Beam: 515.5 + 11.77 × (10 − 3.93)^2 = 949 in^4 (39,500.4 cm^4); $I = 187 + 949 = 1136$ in^4 (47,283.9 cm^4); $S_c = 1136/3.93 = 289.1$ in^3 (4737.5 cm^3); $S_{bs} = 1136/14.07 = 80.7$ in^3 (1322.4 cm^3).

3. Transform the composite section in the region of negative moment to an equivalent section of steel; compute the section moduli

Referring to Fig. 58b, we see that the transformed width = 11/9 = 1.22 in. (31.0 mm). Take static moments with respect to the neutral axis. Or, 11.77(10 − y) − ½(1.22y^2) = 0; $y = 7.26$ in. (184.4 mm). Compute the moment of inertia. Thus, slab: (1/3)1.22(7.26)^3 = 155.6 in^4 (6476.6 cm^4). Beam: 515.5 + 11.77(10 − 7.26)^2 = 603.9 in^4 (25,136.2 cm^4); $I = 155.6 + 603.9 = 759.5$ in^4 (31,612.8 cm^4). Then $S_c = 759.5/7.26 = 104.6$ in^3 (1714.1 cm^3); $S_{ts} = 759.5/10.74 = 70.7$ in^3 (1158.6 cm^3).

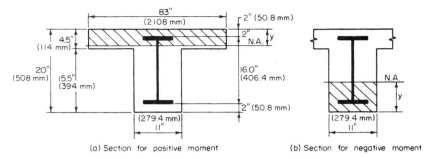

FIGURE 58. Steel beam encased in concrete. (*a*) Section for positive moment; (*b*) section for negative moment.

4. Compute the bending stresses at midspan
The loads carried by the noncomposite member are: slab, (4.5)150(10)/12 = 563 lb/lin ft (8.22 kN/m); stem, 11(15.5)150/144 = 178 lb/lin ft (2.6 kN/m); steel, 40 lb/lin ft (0.58 kN/m); total = 563 + 178 + 40 = 781 lb/lin ft (11.4 kN/m). The load carried by the composite member = 145(10) = 1450 lb/lin ft (21.2 kN/m). Then M^n = (1/8)781(28.5)^{2}12 = 951,500 in.·lb (107.5 kN·m); M^c = (1/20)1450(28.5)^{2}12 = 706,600 in.·lb (79.8 kN·m); f_c = 706,600/[289.1(9)] = 272 lb/sq.in. (1875 kPa), which is acceptable. Also, f_{bs} = (951,500/64.4) + (706,600/80.7) = 23,530 lb/sq.in. (162.2 MPa), which is acceptable.

5. Compute the bending stresses at the support
Thus, M^c = 706,600(20/12) = 1,177,700 in.·lb (132.9 kN·m); f_c = 1,177,700/[104.6(9)] = 1251 lb/sq.in. (8.62 MPa), which is satisfactory. Also, f_{ts} = 1,177,700/70.7 = 16,600 lb/sq.in. (114.9 MPa), which is acceptable. The design is therefore satisfactory with respect to flexure.

6. Investigate the composite member with respect to horizontal shear in the concrete at the section of contraflexure
Assume that this section lies at a distance of 0.2L from the support. The shear at this section is V^c = 1450(0.3)(28.5) = 12,400 lb (55.2 kN).

Refer to Sec. 1. Where the bending moment is positive, the critical plane for horizontal shear is considered to be the surface *abcd* in Fig. 59*a*. For simplicity, however, compute the shear flow at the neutral axis. Apply the relation $q = VQ/I$, where Q = ½(9.22)(3.93)^2 = 71.20 in^3 (1166.8 cm^3) and q = 12,400(71.20)/1136 = 777 lb/lin in. (136 N/mm).

Resistance to shear flow is provided by the bond between the steel and concrete along *bc* and by the pure-shear strength of the concrete along *ab* and *cd*. (The term *pure shear* is used to distinguish this from shear that is used as a measure of diagonal tension.) The allowable stresses in bond and pure shear are usually taken as $0.03f_c'$ and $0.12f_c'$, respectively. Thus bc = 7 in. (177.8 mm); ab = (2.52 + 2^2)$^{0.5}$ = 3.2 in. (81.3 mm); q_{allow} = 7(90) + 2(3.2)360 = 2934 lb/lin in. (419 N/mm), which is satisfactory.

7. Investigate the composite member with respect to horizontal shear in the concrete at the support
The critical plane for horizontal shear is *ef* in Fig. 59*b*. Thus V^c = 1450(0.5)28.5 = 20,660 lb (91.9 kN); Q = 1.22(2)(7.26 − 1) = 15.27 in^3 (250.2 cm^3); q = 20,660(15.27)/759.5 = 415 lb/lin in. (72.7 N/mm); q_{allow} = 7(90) + 2(2)360 = 2070 lb/lin in. (363 N/mm), which is satisfactory.

Mechanical shear connectors are not required to obtain composite action, but the beam is wrapped with wire mesh.

COMPOSITE STEEL-AND-CONCRETE BEAM

A concrete floor slab is to be supported by steel beams spaced 11ft (3.35 m) on centers and having a span of 36 ft (10.97 m). The beams will be supplied with shear connectors to obtain composite action of the steel and concrete. The slab will be 5 in. (127 mm) thick and made of 3000-lb/sq.in. (20.7-MPa) concrete. Loading includes the following: live load, 200 lb/sq.ft. (9.58 kPa); finished floor, ceiling, and partition, 30 lb/sq.ft. (1.44 kPa). In addition, each girder will carry a dead load of 10 kips (44.5 kN) applied as a concentrated load at midspan prior to hardening of the concrete. Conditions at the job site preclude the use of temporary shoring. Design the interior girders, limiting the overall depth of steel to 20 in. (508 mm), if possible.

Calculation Procedure:

1. Compute the unit loads w_1, w_2, and w_3
Refer to the AISC *Specification* and *Manual*. Although ostensibly we apply the elastic-stress method, the design of a composite steel-and-concrete beam in reality is based on the ultimate-strength behavior of the member. Loads that are present before the concrete has hardened are supported by the steel member alone; loads that are present after the concrete has hardened are considered to be supported by the composite member, regardless of whether these loads originated before or after hardening. The effects of plastic flow are disregarded.

The subscripts 1, 2, and 3 refer, respectively, to dead loads applied before hardening of the concrete, dead loads applied after hardening of the concrete, and live loads. The subscripts b, ts, and tc refer to the bottom of the member, top of the steel, and top of the concrete, respectively. The superscripts c and n refer to the composite and noncomposite member, respectively.

We compute the unit loads for a slab weight of 63 lb/lin ft (0.92 kN/m) and an assumed steel weight of 80 lb/lin ft (1167.5 N/m): $w_1 = 63(11) + 80 = 773$ lb/lin ft (11.3 kN/m); $w_2 = 30(11) = 330$ lb/lin ft (4.8 kN/m); $w_3 = 200(11) = 2200$ lb/lin ft (32.1 kN/m).

2. Compute all bending moments required in the design
Thus, $M_1 = 12[(\frac{1}{8})0.773(36)^2 + \frac{1}{4}(10)36] = 2583$ in.·kips (291.8 kN·m); $M_2 = (\frac{1}{8})0.330(36)^2 12 = 642$ in.·kips (72.5 kN·m). $M_3 = (\frac{1}{8})2.200(36)^2 12 = 4277$ in.·kips (483.2 kN·m); $M^c = 2583 + 642 + 4277 = 7502$ in.·kips (847.6 kN·m); $M^n = 2583$ in.·kips (291.8 kN·m); $M_{DL} = 2583 + 642 = 3225$ in.·kips (364.4 kN·m); $M_{LL} = 4277$ in.·kips (483.2 kN·m).

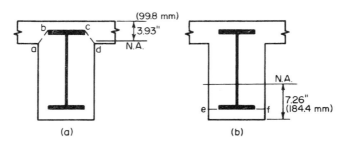

FIGURE 59. Critical planes for horizontal shear.

3. Compute the required section moduli with respect to the steel, using an allowable bending stress of 24 kips/sq.in. (165.5 MPa)

In the composite member, the maximum steel stress occurs at the bottom; in the noncomposite member, it occurs at the top of the steel if a bottom-flange cover plate is used.

Thus, composite section, $S_b = 7502/24 = 312.6$ in^3 (5122.6 cm^3); noncomposite section, $S_{ts} = 2583/24 = 107.6$ in^3 (1763.3 cm^3).

4. Select a trial section by tentatively assuming that the composite-design tables in the AISC Manual are applicable

The *Manual* shows that a composite section consisting of a 5-in. (127-mm) concrete slab, a W18 × 55 steel beam, and a cover plate having an area of 9 sq.in. (58.1 cm^2) provides $S_b = 317.5$ in^3 (5202.9 cm^3). The noncomposite section provides $S_{ts} = 113.7$ in^3 (1863.2 cm^3).

Since unshored construction is to be used, the section must conform with the *Manual* equation $1.35 + 0.35 M_{LL}/M_{DL} = 1.35 + 0.35(4277/3225) = 1.81$. And $S_b^c/S_b^n = 317.5/213.6 = 1.49$, which is satisfactory.

The flange width of the W18 × 55 is 7.53 in. (191.3 mm). The minimum allowable distance between the edge of the cover plate and the edge of the beam flange equals the size of the fillet weld plus $5/16$ in. (7.9 mm). Use a 9 × 1 in. (229 × 25 mm) plate. The steel section therefore coincides with that presented in the AISC *Manual*, which has a cover plate thickness t_p of 1 in. (25.4 mm). The trial section is therefore W18 × 55; cover plate is 9 × 1 in. (229 × 25 mm).

5. Check the trial section

The AISC composite-design tables are constructed by assuming that the effective flange width of the member equals 16 times the slab thickness plus the flange width of the steel. In the present instance, the effective flange width, as governed by the AISC, is $\frac{1}{4}L = \frac{1}{4}(36)12 = 108$ in. (2743 mm); spacing of beams = 132 in. (3353 mm); $16t + 7.53 = 16(5) + 7.53 = 87.53$ in. (2223.3 mm), which governs.

The cross section properties in the AISC table may be applied. The moment of inertia refers to an equivalent section obtained by transforming the concrete to steel. Refer to Sec. 1. Thus $y_{tc} = 5 + 18.12 + 1 - 16.50 = 7.62$ in. (194 mm); $S_{tc} = I/y_{tc} = 5242/7.62 = 687.9$ in^3 (11,272.7 cm^3). From the ACI Code, $f_c = 1350$ lb/sq.in. (9.31 MPa) and $n = 9$. Then $f_c = M^c/(nS_{tc}) = 7,502,000/[9(687.9)] = 1210$ lb/sq.in. (8.34 MPa), which is satisfactory.

6. Record the relevant properties of the W18 × 55

Thus, $A = 16.19$ sq.in. (104.5 cm^2); $d = 18.12$ in. (460 mm); $I = 890$ in^4 (37,044.6 cm^4); $S = 98.2$ in^3 (1609 cm^3); flange thickness = 0.630 in. (16.0 mm).

7. Compute the section moduli of the composite section where the cover plate is absent

To locate the neutral axis, take static moments with respect to the center of the steel. Thus, transformed flange width = 87.53/9 = 9.726 in. (247.0 mm). Further,

Element	A, sq.in. (cm^2)	y, in. (mm)	Ay, in^3 (cm^3)	Ay^2, in^4 (cm^4)	I_o, in^4 (cm^4)
W18 × 55	16.19 (104.5)	0 (0)	0 (0)	0 (0)	890 (37,044.6)
Slab	48.63 (313.7)	11.56 (294)	562.2 (9,212.8)	6,499 (270,509)	101 (4,203,9)
Total	64.82 (418.2)	....	562.2 (9,212.8)	6,499 (270,509)	991 (41,248.5)

Then $\bar{y} = 562.2/64.82 = 8.67$ in. (220 mm); $I = 6499 + 991 - 64.82(8.67)^2 = 2618$ in^4 (108,969.4 cm^4); $y_b = 9.06 + 8.67 = 17.73$ in. (450 mm); $y_{tc} = 9.06 + 5 - 8.67 = 5.39$ in. (136.9 mm); $S_b = 2618/17.73 = 147.7$ in^3 (2420 cm^3); $S_{tc} = 2618/5.39 = 485.7$ in^3 (7959 cm^3).

8. Verify the value of S_b

Apply the value of the K factor in the AISC table. This factor is defined by $K^2 = 1 - S_b$ without plate/S_b with plate. The S_b value without the plate = $317.5(1 - 0.732) = 148$ in^3 (2425 cm^3), which is satisfactory.

9. Establish the theoretical length of the cover plate

In Fig. 60, let C denote the section at which the cover plate becomes superfluous with respect to flexure. Then, for the composite section, $w = 0.773 + 0.330 + 2.200 = 3.303$ kips/lin ft (48.2 kN/m); $P = 10$ kips (44.5 kN); $M_m = 7502$ in.·kips (847.6 kN·m); $R_a = 64.45$ kips (286.7 kN). The allowable values of M_c are, for concrete, $M_c = 485.7(9)1.35/12 = 491.8$ ft·kips (666.9 kN·m) and, for steel, $M_c = 147.7(24)/12 = 295.4$ ft·kips (400.6 kN·m), which governs. Then $R_a x - \frac{1}{2}wx^2 = 295.4$; $x = 5.30$ ft (1.62 m). The theoretical length = $36 - 2(5.30) = 25.40$ ft (7.74 m).

For the noncomposite section, investigate the stresses at the section C previously located. Thus: $w = 0.773$ kips/lin ft (11.3 kN/m); $P = 10$ kips (44.5 kN); $R_a = 18.91$ kips (84.1 kN); $M_c = 18.91(5.30) - \frac{1}{2}(0.773) \times 5.30^2 = 89.4$ ft·kips (121.2 kN·m); $f_b = 89.4(12)/98.2 = 10.9$ kips/sq.in. (75.1 MPa), which is satisfactory.

10. Determine the axial force F in the cover plate at its end by computing the mean bending stress

Thus $f_{mean} = My_{mean}/I = 295.4(12)916.50 - 0.50/5242 = 10.82$ kips/sq.in. (75.6 MPa); $F = Af_{mean} = 9(10.82) = 97.4$ kips (433.2 kN). Alternatively, calculate F by applying the factor $12Q/I$ recorded in the AISC table. Thus, $F = 12QM/I = 0.33(295.4) = 97.5$ kips (433.7 kN).

11. Design the weld required to develop the cover plate at each end

Use fillet welds of E60 electrodes, placed along the sides but not along the end of the plate. The AISC *Specification* requires a minimum weld of $^5/_{16}$ in. (7.9 mm) for a 1-in. (25.4-mm) plate; the capacity of this weld is 3000 lb/lin in. (525 N/mm). Then, length = 97,400/3000 = 32.5 in. (826 mm). However, the AISC requires that the plate be extended 18 in. (457 mm) beyond the theoretical cutoff point, thus providing 36 in. (914 mm) of weld at each end.

12. Design the intermittent weld

The vertical shear at C is $V_c = R_a - 5.30w = 64.45 - 5.30(3.303) = 46.94$ kips (208.8 kN); $q = VQ/I = 46,940(0.33)/12 = 1290$ lb/lin in. (225.9 N/mm). The AISC calls for a minimum weld length of 1½ in. (3.81 mm). Let s denote the center-to-center spacing. Then $s = 2(1.5)3000/1290 = 7.0$ in. (177.8 mm). The AISC imposes an upper limit of 24 times the thickness of the thinner part joined, or 12 in. (304.8 mm). Thus, $s_{max} = 24(0.63) > 12$ in. (304.8 mm). Use a 7-in. (177.8-mm) spacing at the ends and increase the spacing as the shear diminishes.

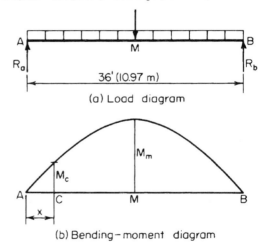

FIGURE 60. (*a*) Load diagram; (*b*) bending-moment diagram.

13. Design the shear connectors

Use ¾-in. (19.1-mm) studs, 3 in. (76.2 mm) high. The design of the connectors is governed by the AISC *Specification*. The capacity of the stud = 11.5 kips (51.2 kN). From the AISC table, V_h = 453.4 kips (2016.7 kN). Total number of studs required = 2(453.4)/11.5 = 80. These are to be equally spaced.

DESIGN OF A CONCRETE JOIST IN A RIBBED FLOOR

The concrete floor of a building will be constructed by using removable steel pans to form a one-way ribbed slab. The loads are: live load, 80 lb/sq.ft. (3.83 kPa); allowance for movable partitions, 20 lb/sq.ft. (0.96 kPa); plastered ceiling, 10 lb/sq.ft. (0.48 kPa); wood floor with sleepers in cinder-concrete fill, 15 lb/sq.ft. (0.72 kPa). The joists will have a clear span of 17 ft (5.2 m) and be continuous over several spans. Design the interior joist by the ultimate-strength method, using f_c' = 3000 lb/sq.in. (20.7 MPa) and f_y = 40,000 lb/sq.in. (275.8 MPa).

Calculation Procedure:

1. Compute the ultimate load carried by the joist

A one-way ribbed floor consists of a concrete slab supported by closely spaced members called ribs, or joists. The joists in turn are supported by steel or concrete girders that frame to columns. Manufacturers' engineering data present the dimensions of steel-pan forms that are available and the average weight of floor corresponding to each form.

Try the cross section shown in Fig. 61, which has an average weight of 54 lb/sq.ft. (2.59 kPa). Although the forms are tapered to facilitate removal, assume for design purposes that the joist has a constant width of 5 in. (127 mm). The design of a ribbed floor is governed by the ACI *Code*. The ultimate-strength design of reinforced-concrete members is covered in Sec. 1.

Referring to the ACI *Code*, compute the ultimate load carried by the joist. Or, w_u = 2.08[1.5(54 + 20 + 10 + 15) + 1.8(80)] = 608 lb/lin ft (8.9 kN/m).

2. Determine whether the joist is adequate with respect to shear

Since the joist is too narrow to permit the use of stirrups, the shearing stress must be limited to the value given in the ACI *Code*. Or, $v_c = 1.1(2\phi)(f_c')(2\phi)^{0.5} = 1.1(2)(0.85)(3000)^{0.5} = 102$ lb/sq.in. (703.2 kPa).

Assume that the reinforcement will consist of no. 4 bars. With ¾ in. (19.1 mm) for fireproofing, as required by the ACI *Code*, d = 8 + 2.5 − 1.0 = 9.5 in. (241.3 mm). The vertical shear at a distance d from the face of the support is V_u = (8.5 − 0.79)608 = 4690 lb (20.9 kN).

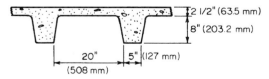

FIGURE 61. Ribbed floor.

The critical shearing stress computed as required by the ACI Code is $v_u = V_u/(bd) = 4690/[5(9.5)] = 99$ lb/sq.in. (682.6 kPa) $< v_c$, which is satisfactory.

3. Compute the ultimate moments to be resisted by the joist
Do this by applying the moment equations given in the ACI Code. Or, $M_{u,pos} = (^1/_{16})608(17)^212 = 132,000$ in.·lb (14.9 kN·m); $M_{u,neg} = (^1/_{11})608(17)212 = 192,000$ in.·lb (21.7 kN·m).

Where the bending moment is positive, the fibers above the neutral axis are in compression, and the joist and tributary slab function in combination to form a T beam. Where the bending moment is negative, the joist functions alone.

4. Determine whether the joist is capable of resisting the negative moment
Use the equation $q_{max} = 0.6375k_187,000/(87,000 + f_y)$, or $q_{max} = 0.6375(0.85)87,000/127,000 - 0.371$. By Eq. 6 of Sec. 1, $M_u = \phi bd^2 f'_c q(1 - 0.59q)$, or $M_u = 0.90(5)9.5^2 \times (3000)0.371(0.781) = 353,000$ in.·lb (39.9 kN·m), which is satisfactory.

5. Compute the area of negative reinforcement
Use Eq. 7 of Sec. 1. Or, $f_c = 0.85(3) = 2.55$ kips/sq.in. (17.6 MPa); $bdf_c = 5(9.5)2.55 = 121.1$; $2bf_c M_u/\phi = 2(5)2.55(192)/0.90 = 5440$; $A_s = [121.1 - (121.1^2 - 5440)^{0.5}]/40 = 0.63$ sq.in. (4.06 cm²).

6. Compute the area of positive reinforcement
Since the stress block lies wholly within the flange, apply Eq. 7 of Sec. 1, with $b = 25$ in. (635 mm). Or, $bdf_c = 605.6$; $2bf_c M_u/\phi = 18,700$; $A_s = [605.6 - (605.6^2 - 18,700)^{0.5}]/40 = 0.39$ sq.in. (2.52 cm²).

7. Select the reinforcing bars and locate the bend points
For positive reinforcement, use two no. 4 bars, one straight and one trussed, to obtain $A_s = 0.40$ sq.in. (2.58 cm²). For negative reinforcement, supplement the two trussed bars over the support with one straight no. 5 bar to obtain $A_s = 0.71$ sq.in. (4.58 cm²).

To locate the bend points of the trussed bars and to investigate the bond stress, follow the method given in Sec. 1.

DESIGN OF A STAIR SLAB

The concrete stair shown in elevation in Fig. 62a, which has been proportioned in conformity with the requirements of the local building code, is to carry a live load of 100 lb/sq.ft. (4.79 kPa). The slab will be poured independently of the supporting members. Design the slab by the working-stress method, using $f'_c = 3000$ lb/sq.in. (20.7 MPa) and $f_s = 20,000$ lb/sq.in. (137.9 MPa).

Calculation Procedure:

1. Compute the unit loads
The working-stress method of designing reinforced-concrete members is presented in Sec. 1. The slab is designed as a simply supported beam having a span equal to the horizontal distance between the center of supports. For convenience, consider a strip of slab having a width of 1 ft (0.3 m).

Assume that the slab will be 5.5 in. (139 mm) thick, the thickness of the stairway slab being measured normal to the soffit. Compute the average vertical depth in Fig. 62b. Thus sec $\theta = 1.25$; $h = 5.5(1.25) + 3.75 = 10.63$ in. (270.0 mm). For the stairway, $w = 100 + 10.63(150)/12 = 233$ lb/lin ft (3.4 kN/m); for the landing, $w = 100 + 5.5(150)/12 = 169$ lb/lin ft (2.5 kN/m).

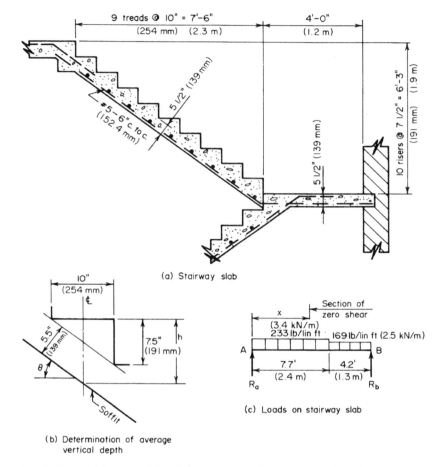

FIGURE 62. (a) Stairway slab; (b) determination of average vertical depth; (c) loads on stairway slab.

2. Compute the maximum bending moment in the slab
Construct the load diagram shown in Fig. 62c, adding about 5 in. (127 mm) to the clear span to obtain the effective span. Thus $R_a = [169(4.2)2.1 + 233(7.7)8.05]/11.9 = 1339$ lb (5.95 kN); $x = 1339/233 = 5.75$ ft (1.75 m); $M_{max} = \frac{1}{2}(1339)5.75(12) = 46,200$ in.·lb (5.2 kN·m).

3. Design the reinforcement
Refer to Table 1 to obtain the following values: $K_b = 223$ lb/sq.in. (1.5 MPa); $j = 0.874$. Assume an effective depth of 4.5 in. (114.3 mm). By Eq. 31, the moment capacity of the member at balanced design is $M_b = K_b b d^2 = 223(12)4.5^2 = 54,190$ in.·lb (6.1 kN·m). The steel is therefore stressed to capacity. (Upon investigation, a 5-in. (127-mm) slab is found to be inadequate.) By Eq. 25, $A_s = M/(f_s j d) = 46,200/[20,000(0.874)4.5] = 0.587$ sq.in. (3.79 cm²).

Use no. 5 bars, 6 in. (152.4 mm) on centers, to obtain $A_s = 0.62$ sq.in. (4.0 cm^2). In addition, place one no. 5 bar transversely under each tread to assist in distributing the load and to serve as temperature reinforcement. Since the slab is poured independently of the supporting members, it is necessary to furnish dowels at the construction joints.

MAXIMUM AVAILABLE MOMENT IN COMPOSITE STEEL AND CONCRETE BEAM

The W24 × 62 beam of Grade SO steel, shown in Fig. 63, is fully encased in concrete and is fabricated without shear anchors. Determine the maximum available moment that the beam can support under LRFD and ASD calculations.

Calculation Procedure:

1. Determine the plastic modulus of the W24 × 62 beam
Using the AISC *Manual* Table 3-2, the plastic modulus of the W24 × 62 beam is:

$$Z_x = 153 \text{ in.}^3 \text{ (0.00249 cu m)}$$

2. Compute the beam flexural strength from
Using the moment relation

$$M_n = Z_x F_y$$
$$= 153 \times \frac{50}{12}$$
$$= 638 \text{ kip-ft (86.44 kNm)}$$

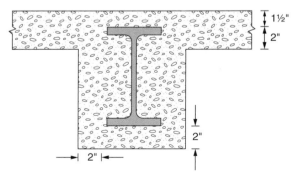

FIGURE 63. Details of encased composite beam.

3. Find the maximum available moment under LRFD and ASD
Use the appropriate relation for each method of analysis. The flexural capacity of this beam is given by AISC 360, Sec. I3.3 as

LRFD	ASD
$\phi_b M_n = 0.90 \times 638$	$M_n/\Omega_b = 638/1.67$
$= 574$ kilo-ft	$= 382.$ kilo-ft
(77.77 kNm)	(51.76 kNm)

Related calculations: For encased composite beams the following comments are important

The concrete encasement effectively inhibits local and flexural buckling of the steel section and AISC 360 Sec. I3.3 permits the following three alternative design methods for determination of the nominal flexural strength:

- The superposition of elastic stresses on the composite section, based on the first yield of the tension flange. Shear anchors are not required as natural bond is considered adequate to connect the steel beam to the concrete.
- The plastic moment strength of the steel section alone. Shear anchors are not required.
- The strength of the composite section obtained from the plastic stress distribution method or the strain-compatibility method. Shear anchors are required to ensure composite action after the breakdown of natural bond.

The available flexural capacity is obtained from AISC 360, Sec. I3.3 as

$$\varphi_b M_n = \text{design flexural capacity}$$
$$= 0.90 M_n$$
$$\frac{M_n}{\Omega_b} = \text{allowable flexural capacity}$$
$$= \frac{M_n}{1.67}$$

This procedure is the work of Alan Williams, Ph.D., S.E., E.I.C.E., C. Eng., as reported in his excellent book *Steel Structures Design* McGraw-Hill, 2011. Wording in the calculation steps and SI values were added by the handbook editor.

INDEX

Note: Page numbers followed by "ff." indicate that the discussion continues on following pages.

Activated sludge reactor, **8.**1 to **8.**8
  aeration tank balance, **8.**5, **8.**6
  biochemical oxygen demand (BOD), **8.**1 to **8.**3, **8.**44 ff.
  food to microorganism ratio, **8.**4
  hydraulic retention time, **8.**2
  oxygen requirements, **8.**3
  reactor volume, **8.**4
  return activate sludge, **8.**4
  sludge quantity wasted, **8.**2
  volume of suspended solids (VSS), **8.**2
  wasted-activated sludge, **8.**4 to **8.**7
Active earth pressure on retaining wall, **4.**36
Aerated grit chamber design, **8.**16 ff.
  air supply required, **8.**18
  grit chamber dimensions, **8.**18
  grit chamber volume, **8.**16
  quantity of grit expected, **8.**18
Aerobic digester design, **8.**12 to **8.**16
  daily volume of sludge, **8.**13
  digester volume, **8.**15
  oxygen and air requirements, **8.**14 to **8.**15
  required VSS reduction, **8.**14
  volume of digested sludge, **8.**14
Air-lift pumps, selection of, **7.**9 to **7.**11
  disadvantages of, **7.**11
Air testing of sewers, **7.**37
Allowable-stress design, **1.**38
Alternative proposals, cost comparisons of, **9.**12 to **9.**22
Altitude of star, **5.**11 to **5.**13
Anaerobic digester design, **8.**44
  BOD entering digester, **8.**44
  daily quantity of volatile solids, **8.**44

Anaerobic digester design (*Cont*):
  percent stabilization, **8.**44
  required volume and loading, **8.**45
  volume of methane produced, **8.**46
Analysis of business operations, **9.**34 to **9.**39
  project planning using CPM/PERT, **9.**34 ff.
Area of tract, **5.**4 to **5.**8
Astronomy, field, **5.**11 to **5.**13
Average-end-area method, **5.**10, **5.**11
Average-grade method, **5.**25 to **5.**28
Axial load:
  and bending, **1.**62
  deformation caused by, **1.**62
  notational system for, **1.**62
  in steel beam column, **1.**34
    in braced frame, **1.**49
  in steel hanger, **1.**62
  stress caused by impact load, **1.**66
  (see also Stress(es) and strain: axial)
Axial member, design load in, **1.**33
Axial shortening of loaded column, **1.**40
Azimuth of star, **5.**11 to **5.**13

Balanced design:
  of prestressed-concrete beam, **2.**51
  of reinforced-concrete beam, **2.**3, **2.**18 ff.
  of reinforced-concrete column, **2.**35
Basins, rapid-mix and flocculation, **8.**28
Beam(s):
  bearing plates for, **1.**59 to **1.**61
  bending stress in, **1.**95 ff.
    and axial force, **1.**13
    with intermittent lateral support, **1.**10
  jointly supporting a load, **1.**96

Beam(s), bending stress in (*Cont*):
    with reduced allowable stress, **1.**11
    in riveted plate girder, **1.**21
    in welded plate girder, **1.**19
    (see also Bending moment)
  biaxial bending, **1.**49 ff.
  column (see Beam column)
  compact and non-compact, **1.**39 ff.
  composite action of, **1.**57
  composite steel-and-concrete, **2.**86 ff.
    highway bridge, **5.**35 to **5.**38
  composite steel-and-timber, **1.**80 ff.
  compound, **1.**79
  concentrated load on, **1.**76
  concrete slab, **1.**57
  conjugate, **1.**91
  continuous, **1.**89 to **1.**93
    of prestressed concrete, **2.**73, **5.**32 to **5.**106
    of reinforced concrete, **2.**23 ff.
    of steel, **1.**17, **1.**129
  deflection of, **1.**89 to **1.**93
  design moment, **1.**44
  economic section of, **1.**10
  lateral torsional buckling, **1.**44
  lightest section to support load, **1.**47 ff.
  minor- and major-axis bending in, **1.**42
  on movable supports, **1.**78
  with moving loads, **1.**103 to **1.**112
  orientation of, **1.**47
  prestressed concrete (see Prestressed-concrete beam)
  reinforced concrete (see Reinforced-concrete beam)
  shape, properties of, **1.**9 ff.
  shear, in concrete:
    connectors for, **1.**57
    force, **1.**56
  shear, in steel:
    center of, **1.**83
    flow, **1.**82
    on yielding support, **1.**96
  shear and bending moment in, **1.**76 ff.
  shear strength, **1.**43 to **1.**47
    of beam web, **1.**82
  shearing stress in, **1.**82 ff.
  statically indeterminate, **1.**95 to **1.**98
  steel (see Steel beam)
  stiffener plates for, **1.**43

Beam(s) (*Cont*):
  strength, for minor- and major-axis bending, **1.**42
  theorem of three moments, **1.**97
  timber (see Timber beam)
  vertical shear in, **1.**78
  web, **1.**58
  web stiffeners for, **1.**43
  welded section, **1.**47
  wood (see Timber beam)
Beam column:
  axial load on, **1.**49
  in braced frame, **1.**49 ff.
  and compression member, **2.**31
  first- and second-order moments in, **1.**49
  flexure and compression combined, **1.**49
  pile group as, **1.**88
  soil prism as, **1.**87
  steel, **1.**127 ff.
Beam connection:
  eccentric load on, **1.**118 ff.
    on pile group, **1.**88
    on riveted connection, **1.**118
    on welded connection, **1.**121
  riveted moment, **1.**117
  semirigid, **1.**147
  of truss members, **1.**146
  welded flexible, **1.**151
  welded moment, **1.**154
  welded seated, **1.**153
Beam web, shear strength of, **1.**58
Bearing plate, for beam, **1.**59
Bending flat plate, **1.**85
Bending moment:
  in bridge truss, **1.**108
  in column footing, **2.**42 ff.
  in concrete girders in T-beam bridge, **5.**35
  diagram, **1.**78 ff.
    for beam column, **1.**34
    for combined footing, **2.**42 to **2.**45
    for steel beam, **1.**13 ff.
    for welded plate girder, **1.**22
  in prestressed-concrete beam, **2.**59
  in reinforced-concrete beam, **2.**4
    compression member, **2.**35
  in steel beam, **1.**75 ff.
    continuous, **1.**129

Bending moment, in steel beam (*Cont.*):
  equation for, **1.**97 ff.
  by moment distribution, **1.**76 ff.
  for most economic section, **1.**8
  and shear in, **1.**76 ff.
  on yielding support for, **1.**95
  and theorem of three moments, **1.**97 ff.
  in three-hinged arch, **1.**110
  in welded plate girder, **1.**19 ff.
Bending stress:
  and axial force, **1.**13
  and axial load, **1.**50 ff.
  in beam, **1.**75 ff.
  in curved member, **1.**87
  with intermittent lateral support, **1.**10
  jointly supporting a load, **1.**96
  in riveted plate girder, **1.**21
  in welded plate girder, **1.**22
  (see also Bending moment)
Benefit-cost analysis, **9.**33
Bernoulli's theorem, **6.**7 ff.
Biogas plants, **8.**51
Bolted splice design, **3.**10
Borda's formula for head loss, **6.**14
Boussinesq equation, **4.**6, **4.**7
Bracing, diagonal, **1.**171
Branching pipes, **6.**15 ff.
Bridge, design procedure for, **5.**40 ff.
Bridge(s), highway, **5.**32 to **5.**106
  composite steel-and-concrete beam,
      2.90 to **2.**94
  bridge, **5.**35
  concrete T-beam, **5.**32
  girder, design of, **5.**32
Bridge filter sizing, traveling grate,
    8.23 ff.
Bridge girder, prestressed concrete, design
    of, **5.**45 ff.
  truss, **1.**104, **5.**93 ff.
Bulkhead:
  anchored analysis of, **4.**16
  cantilever analysis of, **4.**17
  thrust on, **4.**16
Buoyancy, **6.**3 to **6.**5
Business operations, analysis of,
    9.34 to **9.**39
  project planning using CPM/PERT,
      9.34 ff.
Butt splice, **1.**15

Cantilever bulkhead analysis, **4.**17
Cantilever retaining wall, **2.**46 ff.
Cantilever wind-stress analysis, **1.**164
Capacity-reduction factor, **2.**4
Capital, recovery of, **9.**6
Cardboard, corrugated, in municipal waste,
    4.34
Cash flow calculations, **9.**9
Celestial sphere, **5.**12
Centrifugal pump(s), **6.**24 to **6.**50
  characteristic curves for series
      installation, **6.**24 ff.
  cost reduction for:
    energy consumption and loss,
        6.60 ff.
    maintenance for, **6.**36
  driver speed, **6.**30 ff.
  energy cost reduction, **6.**59
  as hydraulic turbines, **6.**52
    applications for, **6.**57
    cavitation potential in, **6.**52
    constant-speed curves, **6.**53
    converting turbine to pump
        conditions, **6.**52
    number of stages of, **6.**52
    performance and flow rate, **6.**53
    specific speed of, **6.**52
  maintenance costs, reducing, **6.**37
  materials and parts selection,
      6.36 ff.
  minimum safe flow, **6.**50
  optimal operating speed, **6.**32 ff.
  ratings table for, **6.**36
  reverse flow, prevention of, **6.**27
  for safety-service, **6.**57 ff.
  seals, mechanical, **6.**43
  selection of, **6.**36 ff.
    for any system, **6.**36 ff.
    best operating speed, **6.**32 ff.
    most common error in, **6.**57
    for reduced cost, **6.**60 ff.
    similarity or affinity laws and,
        6.30 ff.
  similarity or affinity laws, **6.**30 ff.
Centrifuge, solid-bowl, for dewatering,
    8.23 ff.
Channel:
  nonuniform flow in, **6.**17
  uniform flow in, **6.**16

Chlorination for wastewater disinfection, **8.**49
  capacity requirements, **8.**49
  coliform reduction in effluent, **8.**49
Circular settling tank design, **8.**8 to **8.**10
  number of tanks, **8.**9
  peak flow, **8.**8
  surface area, **8.**8
Column(s):
  base, **1.**158
    grillage type, **1.**160 ff.
  reinforced-concrete (see Reinforced-concrete column)
  steel (see Steel column)
  timber, **3.**6, **3.**7
Combined bending and axial loading, **1.**85
Communicating vessels, discharge between, **6.**20
Complete bridge structure, design of, **5.**32 ff.
Composite mechanisms, theorem of, **1.**134
Composite member, thermal effects in, **1.**72 ff.
Composite steel-and-concrete beam, **2.**86 ff.
  highway bridge, **5.**54 to **5.**58
Composite steel-and-concrete column, **1.**56
Composite steel-and-timber beam, **1.**80 ff.
Composition of soil, **4.**2
Compound shaft, analysis of, **1.**74
Compression index, **4.**26
Compression member design:
  by ultimate-strength method, **2.**31 ff.
  by working-stress method, **2.**35 ff.
Compression test:
  triaxial, **4.**10
  unconfined, **4.**8 to **4.**10
Concrete, modulus of elasticity of, **1.**55
Concrete, prestressed (see Prestressed-concrete beam(s))
Concrete, reinforced (see Reinforced-concrete beam(s))
Concrete slab, composite action of, **1.**57
Conjugate-beam method, **1.**91

Connection:
  beam-to-column (see Beam connection)
  horizontal shear resisting, **1.**18 ff.
  pipe joint, **1.**116
  riveted, **1.**112 to **1.**118
  semirigid, **1.**147
  timber, **3.**8 to **3.**12
  welded, **1.**151 to **1.**160
Connectors, shear, **1.**57 ff.
Contaminants, soil (see Soil: contaminated)
Continuity, equation of, **6.**8
Converse-Labarre equation, **4.**29
Copper wire, in municipal wastes, **4.**34, **4.**35
Cost, plant capital, estimation of, **9.**66
Cost vs. benefit analysis, **9.**33
Coulomb's theory, **4.**13
Cover plates:
  for highway girder, **5.**32 ff.
  for plate girder, **1.**19 ff.
  for rolled section, **1.**19
Critical depth of fluid flow, **6.**17 ff.
Critical-path method (CPM) in project planning, **9.**34 ff.
Culmination of star, **5.**14
Curve:
  circular, **5.**14 to **5.**18
  compound, **5.**18 to **5.**20
  double meridian, **5.**4, **5.**5
  horizontal, **5.**18
  sight, **5.**31
  transition, **5.**20 to **5.**25
  vertical, **5.**31
Cylindrical shaft, torsion of, **1.**74

Darcy-Weisbach formula, **6.**11
Declination of star, **5.**13
Deflection:
  of beam, **1.**89 to **1.**93
  of cantilever frame, **1.**93 ff.
  conjugate-beam method, **1.**91
  double-integration method, **1.**89
  moment-area method, **1.**90
  under moving loads, **1.**12
  of prestressed-concrete beam, **2.**59

Deflection (*Cont.*):
    of reinforced-concrete beam, **2.**30 ff.
    unit-load method, **1.**92
    virtual, **1.**126
Deformation of built-up member, **1.**126, **1.**131
Degree of saturation, **4.**2
Departure of line, **5.**2
Depreciation and depletion, **9.**9 ff.
    declining-balance, **9.**12
    straight-line, **9.**10 ff.
    sum-of-the-digits, **9.**12
Depth factor, **3.**3
Design, of complete bridge structure, **5.**40 ff.
    double-T roof in prestressed concrete, **5.**93
    post-tensioned girder in prestressed concrete, **5.**40
Dewatering of sludge, **8.**19 to **8.**23
Digester, wastewater sludge:
    aerobic system design, **8.**12 to **8.**16
    anaerobic system design, **8.**44 ff.
Disinfection of wastewater, chlorinated, **8.**49
Displacement of truss joint, **1.**30
Distance:
    double meridian, **5.**4, **5.**5
    sight, **5.**31
Double-integration method, **1.**89
Double-T roof, design of, **5.**93 ff.
Drainage pump, flow through, **6.**9
Drawdown in gravity wells, **7.**1 to **7.**8
    recovery, **7.**6 to **7.**9
Dummy pile, **4.**33
Dupuit formula in gravity well analysis, **7.**2, **7.**4

Earth thrust:
    on bulkhead, **4.**16 to **4.**19
    on retaining wall, **4.**13
    on timbered trench, **4.**14 to **4.**16
Earthwork requirements, **5.**10
Economics, engineering (see Engineering economics)
Effluent (see Sanitary sewer system design)
Elastic design, **1.**122
Elasticity, modulus of, **1.**62 ff.

Electrodialysis, area and power requirements, **8.**55
Embankment, stability of, **4.**20 to **4.**24
Engineering economics, **9.**4 to **9.**67
    alternative proposals, **9.**12 to **9.**22
    annual cost, after-tax basis, **9.**19
    annual cost of asset, **9.**23
    annual-cost studies, **9.**15
    asset replacement, **9.**20 to **9.**22
    capitalized cost, **9.**8 to **9.**20
    cost and income, **9.**14
    minimum asset life, **9.**14
    analysis of business operations, **9.**39 to **9.**42
    project planning using CPM/PERT, **9.**34 ff.
    capital recovery, **9.**6
    depreciation and depletion, **9.**9 to **9.**12
        accelerated cost recovery, **9.**10
        declining-balance, **9.**12
        effects of inflation, **9.**24 to **9.**26
        sinking-fund, **9.**16 ff.
        straight-line, **9.**9 ff.
        sum-of-the-digits, **9.**12
    evaluation of investments, **9.**26 to **9.**33
        allocation of capital, **9.**27 to **9.**30
        benefit-cost analysis, **9.**33
        payback period, **9.**32
        premium worth method, **9.**26
    inflation, effects of, **9.**24 to **9.**26
        anticipated, **9.**25
        at constant rate, **9.**24
        on replacement cost, **9.**24
        at variable rate, **9.**24
    interest calculations, **9.**3 to **9.**9
        compound, **9.**4
        effective rate, **9.**6
        simple, **9.**4
    nonuniform series, **9.**7
    perpetuity determination, **9.**7
    present worth, **9.**5 ff.
        of continuous cash flow of uniform rate, **9.**8
        of single payment, **9.**5
        of uniform-gradient series, **9.**6
        of uniform series, **9.**6
    probability, **9.**42 ff.
    sinking fund, principal in, **9.**5
    sinking-fund deposit, **9.**5

Engineering economics (*Cont.*):
  statistics and probability, **9.**42 ff.
    arithmetic mean and median,
      9.43 ff.
    decision making, **9.**54
    of failure, **9.**57
    forecasting with a Markov process,
      9.63
    Monte Carlo simulation of
      commercial activity, **9.**59
    normal distribution, **9.**49
    number of ways of assigning work,
      9.46
    Pascal distribution, **9.**47
    perpetuity determination, **9.**7
    Poisson distribution, **9.**48
    population mean, **9.**53
    standard deviation, **9.**43 ff.
    standard deviation from regression
      line, **9.**62
    uniform series, **9.**8 ff.
Enlargement of pipe, **6.**13 ff.
Environmental pollution (see Pollution,
  environmental)
Equipotential line, **4.**4
Equivalent-beam method, **4.**18
Euler equation, for column strength,
  1.26
Evaluation of investments, **9.**26 to **9.**39
Eyebar design, **1.**144

Fatigue loading, **1.**38
Field astronomy, **5.**11 to **5.**14
Fixed-end moment, **1.**97 ff.
Flexural analysis:
  allowable-stress design, **1.**122 ff.
  load and resistance factor design, **1.**8 to
    1.61
  ultimate-stress design, **2.**4 ff.
  working-stress design, **2.**18 to **2.**30
Flocculation and rapid-mix basin, **8.**26
Flow line of soil mass, **4.**4
Flow net, **4.**4
Flowing liquid, power of, **6.**10
Fluid flow, **6.**5 to **6.**22
  (see also Pump(s) and piping systems)
Fluid mechanics, **6.**2 to **6.**29
  Francis equation in fluid discharge,
    6.10 ff.

Fluid mechanics (*Cont.*):
  hydrostatics, **6.**2 to **6.**7
  of incompressible fluids, **6.**5 to **6.**22
  incompressible fluids, mechanics of, **6.**5
    to **6.**22
  in pipes, flow determination for,
    6.12
  power of flowing, **6.**10
  raindrop, velocity of, **6.**21
  specific energy of mass of,
    6.17 ff.
Footing:
  combined, **2.**39 to **2.**42
  isolated square, **2.**40
  settlement of, **4.**27
  sizing by Housel's method, **4.**28
  stability of, **4.**24
Force, hydrostatic, **6.**2 to **6.**5
Francis equation for fluid discharge,
  6.10 ff.
Freyssinet cables, **2.**67

General wedge theory, **4.**14 to **4.**16
Girder(s):
  plate, **1.**19 to **1.**22, **5.**32 to **5.**65
  posttensioned, design of, **2.**67
  steel plate, **1.**83 to **1.**85
  T-beam, **2.**7 ff.
  wood-plywood, **3.**4 to **3.**6
Graphical analysis:
  of pile group, **4.**29
Gravity wells, **7.**1 to **7.**9
  applications for, **7.**3
  base pressure curve, **7.**3
  discharging type, **7.**4 to **7.**6
"Green" products, in reducing pollution,
  4.36
Grillage, as column support, **1.**160
Grit chamber, aerated, **8.**16 to
  8.18
  air supply required, **8.**18
  chamber dimensions, **8.**18
  chamber volume, **8.**16
  quantity of grit expected, **8.**18
Groundwater:
  and drawdown of gravity well, **7.**2 to
    7.4, **7.**17
  and ground surface, **7.**3
Gusset plate analysis, **1.**146

Hanger, steel, **1**.145
Head (pumps and piping):
   capacity:
      rotary pump ranges, **6**.40
      variable-speed, **6**.39 ff.
   computation of, **6**.36
   curves:
      plotting for, **6**.44 ff.
      system head, **6**.24
      types of, **6**.27
   effect of change in, **6**.31
   head loss, **6**.13 ff., **7**.11 to **7**.16
      Borda's formula for, **6**.13
      from pipe enlargement, **6**.13 ff.
      table for, **6**.35
   lift vs. friction, **6**.44 ff.
   in parallel pumping, **6**.27
   pressure loss (see head loss, above)
   in pump selection, **6**.36
   for vapor free liquid, **6**.33 ff.
   weir, variation on, **6**.10 ff.
Highway(s):
   bridge, rain runoff, **5**.106
   bridge design, **5**.32 to **5**.106
   transition spiral, **5**.20 to **5**.24
   volume of earthwork, **5**.10
Hoop stress, **1**.70
Hot-liquid pumps, suction head in, **6**.61
Housel's method, **4**.28
Hydraulic gradient, **4**.3, **6**.16
   in quicksand conditions, **4**.3
Hydraulic jump, **6**.18 ff.
   and power loss, **6**.22
Hydraulic radius, **6**.13
Hydraulic similarity, **6**.22
Hydraulic turbines:
   centrifugal pumps as, **6**.52 ff.
      applications for, **6**.56
      cavitation in, **6**.55 ff.
      constant-speed curves, **6**.55
      converting turbine to pump conditions, **6**.53
      number of stages of, **6**.53
      performance and flow rate, **6**.54
      specific speed of, **6**.53
Hydro power, **6**.52 to **6**.68
   "clean" energy from, **6**.65 ff.
   DOE cost estimates for, **6**.63

Hydro power (*Cont.*):
   Francis turbine in, **6**.62 ff.
   generating capacity, **6**.62
   generator selection, **6**.55
   small-scale generating sites, **6**.62 to **6**.65
   tail-water level, **6**.63
   turbines:
      design of, **6**.53 ff.
      efficiency and load sharing, **6**.64
      Francis turbine, **6**.62
      performance, by type, **6**.65 ff.
      selection of, **6**.66
      tube and bulb type, **6**.64
Hydrocarbons, petroleum, cleanup of, **4**.42
Hydroelectric power (see Hydro power)
Hydropneumatic storage tank sizing, **6**.51
Hydrostatics, **6**.2 to **6**.7
   Archimedes principle, **6**.3
   buoyancy and flotation, **6**.2 to **6**.7
   hydrostatic force, **6**.2 to **6**.5
      on curved surface, **6**.5
      on plane surface, **6**.3, **6**.4
      pressure prism, **6**.4
Hydroturbines:
   designs for, **6**.53 ff.
   efficiency and load sharing, **6**.64
   Francis turbine, **6**.62
   performance characteristics, **6**.53
   in small generating sites, **6**.62 ff.
   tube and bulb type, **6**.64

Impact load, axial stress and, **1**.66
Incompressible fluids, mechanics of, **6**.5 to **6**.22
   Bernoulli's theorem, **6**.7 ff.
   equation of continuity, **6**.5
Inflation, effects of, **9**.24 ff.
   anticipated, **9**.25
   at constant rate, **9**.24
   on replacement cost, **9**.24
   at variable rate, **9**.24
Influence line:
   for bridge truss, **1**.103 to **1**.112
   for three-hinged arch, **1**.110
Interaction diagram, **2**.32
Investments, evaluation of, **9**.20 to **9**.39
Irrigation, solar-powered pumps in, **6**.69

I.8 INDEX

Joist(s):
    prestressed-concrete, **2**.89 ff.
    wood, bending stress in, **3**.2

Kern distances, **2**.60
Knee, **1**.155 ff.
Krey *f*-circle method of analysis, **4**.22 to **4**.24

Laminated wood beam, design of, **3**.13
Landfills, **4**.26 ff.
    mining of, **4**.26
Lap joint, welded, **1**.120
Lap splice, **1**.114
Laplace equation, **4**.4, **4**.5
Lateral torsional buckling, **1**.44
Latitude of line, **5**.2
Leveling, differential, **5**.8
Light-gage steel beam:
    with stiffened flange, **1**.173
    with unstiffened steel flange, **1**.172
Linear transformation, principle of, **2**.75 ff.
Liquid(s):
    fluid mechanics, **6**.2 to **6**.22
    Francis equation in fluid discharge, **6**.10
    incompressible fluids, mechanics of, **6**.5 to **6**.22
    in pipes, flow determination for, **6**.12
    power of flowing, **6**.10
    specific energy of mass of, **6**.17 ff.
    viscosity, in pumps and piping systems, **6**.6 ff.
Load and resistance factor design (LRFD), **1**.8 to **1**.61
Looping pipes, discharge of, **6**.14 ff.
LRFD (see Load and resistance factor design)

Magnel diagram, **2**.60 ff.
Manning formula factor, **6**.13 ff.
Markov process in sales forecasting, **9**.63 ff.
Maximum available moment, in composite beam, **2**.92
Maxwell's theorem, **1**.112
Mechanism method of plastic design, **1**.131

Member(s), **1**.75 to **1**.88
    axial, design load in, **1**.33
    composite, thermal effects in, **1**.72 ff.
    compression, beam column and, **2**.31
    compression, bending moment for, **2**.31
    curved, bending stress in, **1**.87
    steel tension, **1**.36
    timber, **3**.8
    truss, **1**.103 to **1**.112
    ultimate-strength design, **2**.31 ff.
    working-stress design, **2**.18 ff.
    (see also Beam(s))
Membrane, for electrodialysis, **8**.55
Meridian of observer, **5**.12, **5**.14
Methane in anaerobic digester, **8**.46
Method:
    allowable-stress design (ASD), **1**.122 to **1**.141
    average-end-area, **5**.10
    average-grade, **5**.25 to **5**.28
    cantilever, for wind-stress analysis, **1**.164
    conjugate-beam, **1**.91
    Hardy Cross network analysis, **7**.15 ff.
    Housel's, **4**.28
    load and resistance factor design (LRFD), **1**.8 to **1**.61
    prismoidal, **5**.10
    of slices, **4**.20
    slope-deflection, **1**.89 to **1**.93
    Swedish, for slope stability analysis, **4**.20 to **4**.22
    tangent-offset, **5**.25 to **5**.28
    ultimate-strength design, **2**.4 ff.
    working-stress design, **2**.18 to **2**.30
Mining landfills, **4**.26
Modulus of elasticity:
    for composite steel-and-concrete column, **1**.56
    for concrete, **1**.57
    for steel, **1**.62
Modulus of rigidity, **1**.74
Mohr's circle of stress, **1**.68, **4**.8 to **4**.10
Moisture content of soil, **4**.2
Moment:
    bending (see Bending moment)
    of inertia, **1**.26 ff.
    on riveted connection, **1**.21
    second-order, **1**.49
Moment-area method, **1**.90

Moment distribution, **1.**99
Monod kinetics, **8.**2
Monte Carlo simulation, **9.**59
Moving-load system:
   on beam, **1.**103 ff.
   on bridge truss, **1.**108 to **1.**110
Municipal wastes, recycle profits in, **4.**34 to **4.**36
   benefits from, **4.**35
   landfill space and, **4.**35
   recyclable materials, **4.**34, **4.**35
     price increase of, **4.**34 to **4.**36
   waste collection programs, **4.**35

Nadir of observer, **5.**12
Neutral axis in composite beam, **1.**46
Neutral point, **1.**106
Newspapers, in municipal wastes, **4.**34
Nonuniform series, **9.**7

Oblique plane, stresses on, **1.**67
Orifice, flow through, **6.**8

Parabolic arc, **2.**71 ff.
   change in grade, **5.**29
   coordinates of, **2.**71
   location of station on, **5.**28
   plotting, **5.**25 to **5.**28
   of prestressed-concrete beam, **2.**72 ff.
   summit, **5.**29
Parallel pumping economics:
   characteristic curves for, **6.**27 ff.
   check-valve location, **6.**30
   number of pumps used, **6.**27 ff.
   operating point for, **6.**28
   potential energy savings from, **6.**27
   system-head curve, **6.**28
Pascal probability distribution, **9.**47
Passive earth pressure on retaining wall, **4.**36
Payback period of investments, **9.**32
Permeability of soil, **4.**4
PERT in project planning, **9.**34 ff.
Pile-driving formula, **4.**28
Pile group:
   as beam column, **1.**88
   capacity of friction piles, **4.**29
   under eccentric load, **1.**88
   load distribution in, **1.**88, **4.**30 to **4.**34

Pipe joint, **1.**116
Pipe(s) and piping, **6.**5 to **6.**22
   Bernoulli's theorem, application of, **6.**7
   Borda's formula for head loss in pipe, **6.**13
   clay, in sewer pipes, **7.**30
   drainage pump, **6.**9
   eductor, **7.**10
   enlargement, effect on head, **6.**9 ff.
   flow of water in pipes, **6.**12 ff.
   industrial pipeline diagram, **6.**37
   looping pipes, discharge of, **6.**14 ff.
   pipe fittings:
     resistance coefficients of, **6.**33
     resistance of, and valves, **6.**34
   pipe size, **6.**13 ff.
     evaluation of, in pump selection, **6.**36
     Manning formula for selection of, **6.**13
     notation for, **6.**5
     for water supply, **7.**12, **7.**17, **7.**25 to **7.**29
   sewer and storm-water, **7.**24 to **7.**36
   suction and discharge piping, **6.**13 ff.
   Venturi meter, flow through, **6.**8
   in water-supply systems, **7.**11 to **7.**21
   (see also Pump(s) and piping systems)
Plant, capital cost estimation, **9.**66
Plastic containers, in municipal waste, **4.**34
Plastic design of steel structures, **1.**122 to **1.**141
   beam column, **1.**34 ff.
   continuous beam, **1.**129
   definitions relating to, **1.**122
   mechanism method of, **1.**126
   rectangular frame, **1.**131 to **1.**134
   shape factors in, **1.**123
   static method of, **1.**124 to **1.**128
   (see also Structural steel engineering and design)
Plastic media trickling filter design, **8.**36 ff.
Plastic modulus, **1.**122
Plastic moment, **1.**122 ff.
Plastification, defined, **1.**122
Plate, bending of, **1.**84 ff.
Plate girder, **1.**13, **5.**32 ff.
Poisson probability distribution, **9.**48
Poisson's ratio, **1.**84 ff.

Pollution, environmental, **4.**34 ff.
  "green" products and, **4.**36
  hydro power, "clean" energy from, **6.**65 ff.
  landfill space, **4.**35
  recycle profit potential, **4.**34 to **4.**36
  waste, municipal, **4.**34 to **4.**36
  waste sites, contaminated, **4.**2
Polymer dilution/feed system sizing, **8.**29
  predictable properties of, **8.**36
  required rotational rate of distributor, **8.**35
  treatability constant, **8.**35
  typical dosing rates for trickling filters, **8.**41
Porosity of soil, **4.**2
Portal method of wind-stress analysis, **1.**162
Post-tensioned girder, design of, **5.**45 ff.
Power of flowing liquid, **6.**10
Present worth:
  of continuous cash flow of uniform rate, **9.**9
  of single payment, **9.**5
  of uniform series, **9.**6
Pressure, soil, **4.**6 to **4.**8
  under dam, **1.**87
Pressure center of hydrostatic force, **6.**3
Pressure prism of fluid, **6.**4
Pressure vessel:
  prestressed, **1.**70
  thick-walled, **1.**70
  thin-walled, **1.**69
Prestress-moment diagram, **2.**75 ff.
Prestressed-concrete beam(s), **2.**51 to **2.**92, **5.**40 ff.
  in balanced design, **2.**50
  bending moment in, **2.**51
  in bridges, **5.**32 ff.
  continuity moment in, **2.**73
  continuous, **2.**73
    with nonprestressed reinforcement, **2.**77 ff.
    reactions for, **2.**89
  deflection of, **2.**62
  girder, **5.**45 ff.
  guides in design of, **2.**59, **5.**32 ff.
  kern of, **2.**60
  linear transformation in, **2.**75 ff.

Prestressed-concrete beam(s) (*Cont.*):
  loads carried by, **2.**59
  Magnel diagram for, **2.**60
  notational system for, **2.**51
  posttensioned, design of, **2.**67 ff.
  posttensioning of, **2.**50
  prestress shear and moment, **2.**51
  pretensioned, design of, **2.**59
  pretensioning of, **2.**51, **2.**59
  radial forces in, **2.**72
  section moduli of, **2.**58
  shear in, **2.**50
  stress diagrams for, **2.**52 ff.
  tendons in, **2.**52, 2,55, **2.**57
  trajectory, force in, **2.**72
  web reinforcement of, **2.**63 ff.
Prestressed concrete bridge design, **5.**32 ff.
Prestressed-concrete joist, **2.**89
Prestressed-concrete stair slab, **2.**90
Principal axis, **1.**27
Principal plane, **1.**67 ff.
Principal stress, **1.**68
Prismoidal method for earthwork, **5.**10
Probability:
  of failure, **9.**57
  normal distribution, **9.**52 ff.
  Pascal distribution, **9.**47
  Poisson distribution, **9.**48
  standard deviation, **9.**43
  standard deviation from regression line, **9.**62
Product of inertia, **1.**27
Profit potential, from recycling municipal waste, **4.**34 to **4.**36
Project planning, **9.**34
Pump(s) and piping systems, **6.**24 to **6.**44
  affinity laws for, **6.**30 ff.
  analysis of characteristic curves, **6.**44 ff.
    different pipe sizes, **6.**44 ff.
    duplex pump capacities, **6.**43
    effect of wear, **6.**49
    significant friction loss and lift, **6.**44
  centrifugal (see Centrifugal pump(s))
  check-valve, to prevent reverse flow, **6.**30
  closed-cycle solar-powered system, **6.**69 ff.
  discharge flow rate, **6.**9

Pump(s) and piping systems (*Cont.*):
　drainage pump, flow through, **6.**9
　duplex plunger type, **6.**44 ff.
　exit loss, **6.**13 ff.
　fittings, resistance coefficients of, **6.**33
　head, **6.**44 ff.
　　analysis of characteristic curves, **6.**44 ff.
　　Borda's formula for head loss, **6.**14
　　curves, plotting for, **6.**44 ff.
　　Darcy-Weisbach formula for friction, **6.**11
　　effect of change in, **6.**31
　head loss and pipe enlargement, **6.**13 ff.
　hydraulic jump, **6.**18 ff.
　　power loss resulting from, **6.**18 ff.
　as hydraulic turbines, **6.**52 ff.
　hydropneumatic storage tank, sizing of, **6.**51
　Manning formula for pipe-size selection, **6.**13
　materials for pump parts, **6.**36 ff.
　minimum safe flow, **6.**50
　mostly lift, little friction, **6.**46
　no lift, all friction head, **6.**46
　optimal operating speed, **6.**32
　parallel pumping economics, **6.**27 ff.
　　characteristic curves for, **6.**27 ff.
　　check-valve location, **6.**30
　　number of pumps used, **6.**29
　　operating point for, **6.**29
　　potential energy savings from, **6.**27
　　system-head curve for, **6.**28
　in parallel system, **6.**27
　pipe-size selection, **6.**13
　pump type, by specific speed, **6.**32
　selection for any system, **6.**36 ff.
　　capacity required, **6.**36
　　characteristics of modern pumps, **6.**40
　　characteristics with diameter varied, **6.**43
　　class and type, **6.**38
　　common error in, **6.**57
　　composite rating chart, **6.**41
　　liquid conditions, **6.**37
　　rating table, **6.**40

Pump(s) and piping systems, selection for any system (*Cont.*):
　　selection guide, **6.**37 ff.
　　suction and discharge piping arrangements, **6.**8, **6.**33
　　total head, **6.**33
　　variable-speed head capacity, **6.**39
　selection for reduced energy cost, **6.**57
　　best efficiency point (BEP) for, **6.**61
　　energy efficiency pump, **6.**60
　　and series pump operation, **6.**24
　　specific speed and, **6.**60 ff.
　　specifications for operation below BEP, **6.**60
　selection of materials and parts, **6.**36, **6.**51
　series installation analysis, **6.**24 ff.
　　characteristic curves for, **6.**44
　　in reducing energy consumption, **6.**43
　　seriesed curve, **6.**24
　similarity and affinity laws, **6.**30 ff.
　for small hydro power installations, **6.**62
　solar-powered system, **6.**69
　　applications of, **6.**69
　　closed-cycle design, **6.**69
　　gas-release rate in, **6.**69
　　Rankine-cycle turbine in, **6.**70
　　solar collectors used for, **6.**69
　specific speed, by pump type, **6.**32
　static suction lift, **6.**34
　suction and discharge piping arrangements, **6.**33
　system-head curve, **6.**26 ff.
　table for head loss determination, **6.**36
　total head for vapor free liquid, **6.**33
　valves, resistance of, **6.**32
　variable-speed head capacity, **6.**39
　variation in, on weir, **6.**10 ff.
　velocity, pressure, and potential, **6.**10

Quicksand conditions determination, **4.**3

Rain, runoff from bridge, **5.**106
Rainfall:
　imperviousness of various surfaces to, **7.**25
　raindrop, velocity of, **6.**21

Rainfall (*Cont.*):
  storm-water runoff, **7.**24, **7.**25, **7.**28, **7.**33 to **7.**36
Rankine's theory, **4.**11
Rapid-mix and flocculation basin design, **8.**28
  volume and power requirements, **8.**28
    for flocculation, **8.**29
    for rapid-mix basin, **8.**28
Rebhann's theorem, **4.**13
Reciprocal deflections, theorem of, **1.**112
Recycle profit potentials, in municipal wastes, **4.**34 to **4.**36
Recycling of municipal waste:
  benefits from, **4.**35
  landfill space and, **4.**35
  profit potential in, **4.**34 to **4.**36
  types of material in:
    copper, **4.**34, **4.**35
    corrugated cardboard, **4.**34
    newspapers, **4.**34
    plastics, **4.**34
    prices of, **4.**34 to **4.**36
  waste collection programs, **4.**35
Redtenbacker's formula, **4.**29
Reinforced-concrete beam, **2.**4 to **2.**13
  in balanced design, **2.**6 ff.
  bond stress in, **2.**13
  with compression reinforcement, **2.**9 ff.
  continuous: deflection of, **2.**30 ff.
    design of, **2.**14 to **2.**16
  equations of, **2.**4 ff.
  failure in, **2.**4
  minimum widths, **2.**4
  with one-way reinforcement, **2.**14
  of rectangular section, **2.**7 ff.
  shearing stress in, **2.**11 ff.
    alternative methods for computing, **2.**11
  of T section, **2.**7 ff.
  transformed section of, **2.**20 ff.
  with two-way reinforcement, **2.**16 to **2.**18
  ultimate-strength design of, **2.**6 ff.
  web reinforcement of, **2.**11 to **2.**13, **2.**24 to **2.**26
  working-stress design of, **2.**18 to **2.**32

Reinforced-concrete column, **2.**35 ff.
  in balanced design, **2.**39
  footing for, **2.**40 ff.
  interaction diagram for, **2.**32 to **2.**34
  ultimate-strength design of, **2.**4 ff.
  working-stress design of, **2.**18 ff.
Retaining wall:
  active and passive earth pressure on, **4.**36
  cantilever, design of, **2.**46
  earth thrust on, **4.**11 to **4.**16
Reynolds number, **6.**5 ff.
Right ascension of star, **5.**12
Rigidity, modulus of, **1.**74
Riveted connection(s), **1.**112 to **1.**118
  capacity of rivet in, **1.**113
  eccentric load on, **1.**118 ff.
  moment on, **1.**113
Roof, double-T, design of, **5.**93 ff.
Rotary-lobe sludge pump sizing, **8.**41
  flow rate required, **8.**41
  head loss in piping system, **8.**42
    multiplication factor, **8.**42
  pump horsepower (kW) required, **8.**43
    installed horsepower (kW), **8.**44
  pump performance curve, **8.**43
  pump selection, **8.**40
Roughness coefficient, **6.**13
Route design, **5.**1 to **5.**39

Sanitary sewer system design, **8.**53 ff.
  design factors, **8.**49
  lateral sewer size, **8.**50
  Manning formula conveyance factor, **8.**47
  required size of main sewer, **8.**48
  sanitary sewage flow rate, **8.**46
  sewer size with infiltration, **8.**48
Series pumping, **6.**24 ff.
  characteristic curves for, **6.**44
  in reducing energy consumption, **6.**22
  seriesed curve, **6.**26 ff.
  system-head curve, **6.**26
Settling tank design, **8.**8 to **8.**10

Sewage-treatment method selection, **8.**53 ff.
  biogas plants, **8.**51
  daily sewage flow rate, **8.**50
  industrial sewage equivalent, **8.**50
  typical efficiencies, **8.**51
  wet processes, **8.**53
Sewer, air testing of, **7.**37
Sewer systems (see Storm-water and sewer systems and Sanitary sewer system design)
Shape factor, **1.**18 ff.
Shear:
  in beam, **1.**75 ff.
  in bridge truss, **1.**106, **5.**50 to **5.**52
  in column footing, **2.**40 ff.
  in concrete slab, for composite action, **1.**157
  of prestressed concrete, **2.**53
  punching, **2.**41, **2.**46
  on riveted connection, **1.**118
  for welded connection, **1.**121 ff.
Shear center, **1.**83
Shear connectors, **1.**57
Shear diagram:
  for beam, **1.**75 ff.
  for combined footing, **2.**42
Shearing stress (see Stress(es) and strain: shearing stress)
Shrink-fit stress, **1.**73
Sight distance, **5.**31
Similarity, hydraulic, **6.**22
Sinking fund, **9.**5
Slenderness ratio, **1.**29
Slices, method of, **4.**20 to **4.**22
Slope, stability of:
  by f-circle method, **4.**22 to **4.**24
  method of slices, **4.**20 to **4.**22
Slope-deflection method of wind-stress analysis, **1.**167 ff.
Sludge, sanitary wastewater treatment of, **8.**1 to **8.**44
  activated sludge reactor design, **8.**1 to **8.**8
  aerated grit chamber design, **8.**16 to **8.**18
  aerobic digester design, **8.**12 to **8.**16
  anaerobic digester design, **8.**44 ff.
  rotary-lobe sludge pump sizing, **8.**41 ff.

Sludge, sanitary wastewater treatment of (*Cont.*):
  solid-bowl centrifuge for dewatering, **8.**23 ff.
  thickening of wasted-activated sludge, **8.**4 to **8.**7
Small hydropower sites, **6.**62 ff.
  "clean" energy from, **6.**65
  DOE operating-cost estimates, **6.**66
  efficiency falloff and load sharing, **6.**64
  Francis turbine in, **6.**62
  importance of tail-water level, **6.**63
  turbine design, **6.**52 ff.
  typical power-generating capacity, **6.**62
Soil:
  composition of, **4.**2
  compression index of, **4.**26
  consolidation of, **4.**25
  contaminated, **4.**2, **7.**33
    and water supply, **7.**23
  flow net in, **4.**4 to **4.**6
  moisture content of, **4.**2, **4.**3
  permeability of, **4.**4
  porosity of, **4.**2
  pressure (see Soil pressure)
  quicksand conditions, **4.**3
  shearing capacity of, **4.**8 to **4.**10
  specific weight of, **4.**3
  thrust on bulkhead, **4.**16
Soil mechanics, **4.**1 to **4.**37
  mining landfills, **4.**26
  municipal wastes, recycle profits in, **4.**34 to **4.**36
Soil pressure:
  caused by point load, **4.**6
  caused by rectangular loading, **4.**7
  under dam, **1.**87

Solar-powered pumps, **6.**69
  applications of, **6.**69
  closed-cycle design, **6.**69
  gas-release rate in, **6.**69
  Rankine-cycle turbine in, **6.**69
  refrigerant selection for, **6.**69
  solar collectors used for, **6.**69

Solid-bowl centrifuge for sludge
    dewatering, **8.**49 ff.
  capacity and number of centrifuges,
    8.19
  centrifugal force, **8.**22
  dewatered sludge cake discharge rate,
    8.21
  selecting number of centrifuges needed,
    8.19
  sludge feed rate required, **8.**20
  solids capture, **8.**15
Space frame, **1.**135
Spiral, transition, **5.**20 to **5.**25
Stability:
  of embankment, **4.**20 to **4.**24
  of footing, **4.**24
  of slope, **4.**20 to **4.**24
  of vessel, **6.**6
Stadia surveying, **5.**9
Stair slab, prestressed-concrete, **2.**90
Star, azimuth of, **5.**11 to **5.**13
  culmination of, **5.**14
Star strut, **1.**28
Statically indeterminate structures, **1.**95 to
    1.101
  beam(s), **1.**95 ff.
    bending moment of, **1.**96
    bending stress of beam, **1.**96
    theorem of three moments, **1.**97
    truss, analysis of, **1.**101
Statistics and probability, **9.**42 ff.
Steel beam(s), **1.**8 to **1.**28
  composite concrete and, **2.**84 ff.
  continuous, **1.**94 ff.
    elastic design of, **1.**122 ff.
    plastic design of, **1.**122
  with continuous lateral support, **1.**8
  cover-plated, **1.**22
  encased in concrete, **2.**84 ff.
  with intermittent lateral support, **1.**10
  light gage, **1.**172 ff.
  with reduced allowable stress, **1.**11 ff.
  shear in, **1.**76 ff.
  shearing stress in, **1.**76 ff.
  stiffener plates for, **1.**121
Steel column, **1.**27 to **1.**35
  axial shortening when loaded, **1.**33
  base for, **1.**158 ff.
  beam column, **1.**34 ff., **1.**49

Steel column (*Cont.*):
  built-up, **1.**27
  compressive strength, **1.**26
    of composite, **1.**55 ff.
    of welded section, **1.**27
  concrete-filled, **1.**55
  effective length of, **1.**29
  with end moments, **1.**34
  under fatigue loading, **1.**33
  with grillage support, **1.**160
  with intermediate loading, **1.**32
  lacing of, **1.**31
  with partial restraint, **1.**30
  of star-strut section, **1.**28
  with two effective lengths, **1.**29
  welded section, **1.**121
Steel hanger analysis, **1.**145
Steel structures (see Structural steel
    engineering and design)
Steel tension member, **1.**34
Stiffener plates, **1.**43, **1.**151 to **1.**154
Storage tank, hydropneumatic, **6.**51
Storm-water and sewer systems, **7.**24 to
    7.36
  runoff rate and rainfall intensity, **7.**24
    by area, **7.**25
    rational method, **7.**24
    by surface, **7.**24
    Talbot formulas for, **7.**24
    (see also Sanitary sewer system
      design)
  sewer pipes, **7.**25 to **7.**36
    bedding requirements of, **7.**29 to **7.**33
    capacities, **7.**34
    clay pipe strength, **7.**30
    earth load on, **7.**29 to **7.**33
    embedding method, selection of, **7.**31
    sanitary systems, **7.**28
    separate vs. combined design types,
      7.36
    sizing for flow rates, **7.**25 to **7.**29, **7.**33
    slope of, **7.**26, **7.**36
    typical plot plan, **7.**35
Stress(es) and strain, **1.**62 to **1.**74
  axial, **1.**62 to **1.**63, **1.**158
  bending, **1.**76 ff.
  bond, **2.**13 ff.
  in compound shaft, **1.**74
  in flexural members, **1.**75 ff.

Stress(es) and strain (*Cont.*):
  hoop, **1.**69
  moving loads, **1.**103 to **1.**112
  on oblique plane, **1.**67
  principal, **1.**68
  in rectangular beam, **2.**20
  shearing stress, **1.**62 ff.
    cylindrical, torsion of, **1.**74
    in homogeneous beam, **1.**80
    in prestressed cylinder, **1.**70
    in reinforced-concrete beam, **2.**9 ff.
    shrink-fit and radial pressure, **1.**73
    in steel beam, **1.**76 to **1.**83
    thermal, **1.**72 ff.
    in timber beam, **3.**2, **3.**3
Structural steel engineering and design, **1.**1 to **1.**175
  axial member, design load in, **1.**62
  beam connection:
    riveted moment, **1.**149 to **1.**151
    semirigid, **1.**151
    welded moment, **1.**154
    welded seated, **1.**153
  column base, for axial load, **1.**158
    for end moment, **1.**158
    grillage type, **1.**160
  composite steel-and-concrete beam, **1.**80, **2.**92
  bridge, **5.**35 ff.
  connection, beam-to-column:
    of truss members, **1.**146
  eccentric load, **1.**118
    on pile group. **1.**88
    on rectangular section, **1.**95
    on riveted connection, **1.**118
    on welded connection, **1.**121
  eyebar, **1.**144
  gusset plate, **1.** 146
  hanger, steel, **1.**145
  knee:
    curved, **1.**155 to **1.**157
    rectangular, **1.**155
  stair slab, **2.**90
  steel beam, **1.**172 to **1.**173
    encased in concrete, **2.**84 ff.
    light-gage, **1.**172 to **1.**173
  wind drift, **1.**169 to **1.**171
    reduction with diagonal bracing, **1.**171

Structural steel engineering and design (*Cont.*):
  wind-stress analysis, **1.**164 to **1.**167
    cantilever method, **1.**164
    portal method, **1.**162
    slope-deflection method, **1.**167
Surveying, **5.**1 to **5.**31
  field astronomy, **5.**11 to **5.**14
  land and highway, **5.**1 to **5.**31
  stadia, **5.**9
Swedish method for slope analysis, **4.**20 to **4.**22

T-beam reinforced-concrete, **1.**173, **2.**7 ff.
Tangent-offset method, **5.**28
Tangential deviation, **1.**90
Tank(s):
  aeration, in wastewater treatment, **8.**6
  circular, in wastewater treatment, **8.**8 to **8.**10
  hydropneumatic, **6.**57
Temperature reinforcement, **1.**173
Tension member, steel, **1.**26
Terzaghi general wedge theory, **4.**14
Terzaghi theory of consolidation, **4.**25
Thermal effects in structural members, **1.**71 ff.
Thermodynamics, first law of, **6.**35
Three moments, theorem of, **1.**97 ff.
Timber beam, **3.**1 to **3.**4
  bending stress in, **3.**2
  bolted splice, **3.**10
  composite steel and, **1.**45 to **1.**47
  depth factor of, **3.**3
  lateral load on nails in, **3.**9
  screw loads in, **3.**10
  shearing stress in, **3.**3
Timber column, **3.**6, **3.**7
Timber connection, **3.**11
Timber engineering, **3.**1 to **3.**13
Timber member under oblique force, **3.**7
Torsion of shaft, **1.**74
Tract:
  area of, **5.**5 to **5.**8
    irregular, **5.**7
    rectilinear, **5.**4
  partition of, **5.**5 to **5.**7

Trajectory:
  concordant, **2.**77
  linear transformation of, **2.**75 ff.
Transformed section, **1.**80
Transition spiral, **5.**20 to **5.**25
Traveling-grate bridge filter sizing, **8.**23 to **8.**25
Traverse, closed, **5.**2 to **5.**7
Trench, earth thrust on timbered, **4.**4 to **4.**16
Trickling filter design, **8.**29 to **8.**33
  BOD loading for first-stage filter, **8.**31
  filter efficiency, **8.**29
  plastic media type, **8.**33 to **8.**36
Truss:
  bridge, **1.**104, **5.**50
  influence line:
    for bending moment in, **1.**108
    with moving loads, **1.**106 ff.
    for shear in, **1.**104
    statically indeterminate, **1.**101
    by uniform loads, **1.**106
  joint, displacement of, **1.**65
Turbines:
  hydraulic:
    centrifugal pumps as, **6.**52 ff.
    converting turbine to pump conditions, **6.**52
  hydroturbines:
    designs for, **6.**52 to **6.**58
    efficiency and load sharing, **6.**52
    Francis turbine, **6.**62
    performance, by type, **6.**65
    in small-scale generating sites, **6.**62
    tube and bulb type, **6.**62 ff.
Turbulent flow, **6.**11 ff.
Two-way slab, **2.**16 ff.

Ultimate load, **1.**122
Ultimate-strength design:
  for compression members, **2.**31 ff.
  for flexural members, **2.**4 ff.
Uniform series, **9.**6 ff.
Unit-load method, **1.**92

Venturi meter, flow through, **6.**8 ff.
Vertical parabolic curve:
  containing given point, **5.**29

Vertical parabolic curve (*Cont.*):
  plotting of, **5.**25 to **5.**28
  sight distance on, **5.**31
Virtual displacements, theorem of, **1.**139
Visibility, on vertical curve, **5.**31
Void ratio of soil, **4.**26

Wall, retaining (see Retaining wall)
Waste:
  contaminated sites, **4.**34
  municipal, **4.**34
    incineration of, **4.**35
    landfill area required for, **4.**36
    rate of generation of, **4.**36
    recycle profit potential in, **4.**34 to **4.**36
Waste-activated sludge thickening, **8.**10 to **8.**12
  size of gravity belt thickener, **8.**11
  sludge and filtrate flow rates, **8.**11
  solids capture, **8.**12
Wastewater disinfection, chlorination system for, **8.**44
Wastewater treatment and control, **8.**1 to **8.**55
  (see also Sanitary sewer system design)
Water pollution:
  and hazardous wastes, **7.**23
  impurities in water-supply, **7.**21 to **7.**24
Water-supply systems, **7.**1 to **7.**21
  air-lift pump selection, **7.**9 to **7.**11
    compressor capacity, **7.**9
    submergence, effect of, **7.**10
  choice of pipe for, **7.**11
  demand curve for typical week, **7.**4
  fire safety requirements, **7.**12 ff.
  flow rate and pressure loss, **7.**12 ff.
  flow rates in, **7.**11 ff.
    for domestic water, **7.**12
    friction head loss, **7.**12
    load factor determination, **7.**12
    Hardy Cross network analysis method, **7.**15 ff.
    industrial water and steam requirements, **7.**20
  municipal water sources, **7.**17
  parallel and single piping, **7.**11
  pressure loss analysis, **7.**11 to **7.**16

Water-supply systems (*Cont.*):
  pump drawdown analysis, **7.1** to **7.9**
    drawdown in gravity well, **7.4** to **7.9**
    Dupuit formula in, **7.2**, **7.4**
    recovery-curve calculation, **7.6** to **7.9**
    wells in extended use, **7.6** to **7.9**
  selection of, **7.17** to **7.21**
    fire hydrant requirements, **7.19**, **7.20**
    flow rate computation, **7.17**
    piping for, **7.17**
    pressurizing methods, **7.19**
    water supply sources in, **7.17**
  treatment methods, **7.21** to **7.24**
    disinfection, **7.23**
    filtration, **7.21**
    softening, **7.22**
    solvents, **7.23**
  typical municipal water sources, **7.17**
  water wells, **7.1** to **7.11**
    air-lift pump for, **7.9** to **7.11**
Web reinforcement:
  of prestressed-concrete beam, **2.59**
  of reinforced-concrete beam, **2.9** ff.
Web stiffeners, **1.43**
Wedge of immersion, **6.4**

Weir:
  discharge over, **6.10** ff.
  variation in head on, **6.10**
Welded beams, **1.41** ff.
  design moment of, **1.43**
Welded connection, **1.43**, **1.120** ff.
  welded flexible, **1.151**
  welded moment, **1.154**
  welded seated, **1.154**
Welded plate girder design, **1.19** ff.
Westergaard construction, **4.32**
Williott displacement diagram, **1.65**
Wind drift reduction, **1.169** ff.
Wind stress analysis, **1.162** ff.
  cantilever method, **1.164**
  portal method, **1.162**
  slope-deflection method, **1.167**
Wood beam (see Timber beam)
Wood joist(s), **3.2**
Wood-plywood girder, **3.4** to **3.6**
Working-stress design, **2.35** to **2.39**

Yield moment, **1.122**
Yield-point stress, **1.122**

Zenith of observer, **5.12**, **5.13**